Autodesk Naviswork 2024 for Novices

(Learn by Doing)

CADSoft Technologies

CADSoft Excellence de Technology

CADSoft Technologies

Autodesk Naviswork 2024 for Novices: Learn By Doing

CADSoft Technologies

ISBN 978-1-942689-61-4

NOTICE TO THE READER

Publisher does not warrant or guarantee any of the products described in the text or perform any independent analysis in connection with any of the product information contained in the text. Publisher does not assume, and expressly disclaims, any obligation to obtain and include information other than that provided to it by the manufacturer.

The reader is expressly warned to consider and adopt all safety precautions that might be indicated by the activities herein and to avoid all potential hazards. By following the instructions contained herein, the reader willingly assumes all risks in connection with such instructions.

The publisher makes no representation or warranties of any kind, including but not limited to, the warranties of fitness for particular purpose or merchantability, nor are any such representations implied with respect to the material set forth herein, and the publisher takes no responsibility with respect to such material. The publisher shall not be liable for any special, consequential, or exemplary damages resulting, in whole or part, from the reader's use of, or reliance upon this material.

www.cadsofttech.com

DEDICATION

To students, who are dedicated to learning new technologies

THANKS

To employees of CADSoft Technologies
for their cooperation

Online Training Program Offered by CADSoft Technologies

CADSoft Technologies provides effective and affordable virtual online training on various software packages including Computer Aided Design, Manufacturing and Engineering (CAD/CAM/CAE), computer programming languages, animation, architecture, and GIS. The training is delivered 'live' via Internet at any time, any place, and at any pace to individuals as well as the students of colleges, universities, and CAD/CAM/CAE training centers. The main features of this program are:

Training for Students and Companies in a Classroom Setting

Highly experienced instructors and qualified Engineers at CADSoft Technologies conduct the classes under the guidance of Prof. Sham Tickoo of Purdue University Northwest, USA. This team has authored several textbooks that are rated "one of the best" in their categories and are used in various colleges, universities, and training centers in North America, Europe, and in other parts of the world.

Training for Individuals

CADSoft Technologies with its cost effective and time saving initiative strives to deliver the training in the comfort of your home or work place, thereby relieving you from the hassles of traveling to training centers.

Training Offered on Software Packages

CADSoft Technologies provides basic and advanced training on the following software packages:

***CAD/CAM/CAE**: CATIA, Creo Parameteric, SOLIDWORKS, Autodesk Inventor, Solid Edge, NX, AutoCAD, AutoCAD LT, Customizing AutoCAD, AutoCAD Electrical, and ANSYS*

***Architecture and GIS**: Autodesk Revit (Architecture/Structure/MEP), AutoCAD Civil 3D, AutoCAD Map 3D, Autodesk Navisworks, Bentley Staad.Pro, Oracle Primavera P6, MS Project, ArcGIS, and Raster Design.*

***Animation and Styling**: Autodesk 3ds Max, 3ds Max Design, Autodesk Maya, Autodesk Alias, MZBrush, and MAXON CINEMA 4D*

***Computer Programming**: C++, VB.NET, Oracle, AJAX, and Java*

For more information, please visit the following link:
https://www.cadsofttech.com

Table of Contents

Dedication iii
Preface xiii

Chapter 1: Introduction to Autodesk Navisworks 2024

Introduction 1-2
Basic Features of Navisworks Manage 1-2
Benefits of Navisworks 1-2
Starting Autodesk Navisworks 2024 1-2
User Interface 1-3
Title Bar 1-3
Quick Access Toolbar 1-4
Info Centre 1-4
Application Menu 1-5
Ribbon 1-5
Scene View 1-6
Navigation Bar 1-6
ViewCube 1-6
Dockable Windows 1-6
Status Bar 1-7
Autodesk Navisworks Dialog Boxes 1-7
Configuring Settings in Navisworks 1-7
Configuring the Local Settings 1-8
Configuring the Global Settings 1-8
Keyboard Shortcuts 1-8
File Types in Navisworks 1-9
File Readers 1-9
Managing Files in Navisworks 1-9
Working with Batch Utility 1-10
Creating a List of Design Files 1-10
Sharing Data in Navisworks 1-10
Printing Files 1-10
Importing Files 1-10
Exporting Files 1-10
Exporting Images and Animations 1-10
Export Data 1-10
Setting Units in Navisworks 1-10
Navisworks Workspaces 1-11
Safe Mode 1-11
Navisworks Extended 1-11
Navisworks Standard 1-11
Navisworks Minimal 1-11
Autodesk Navisworks Help System 1-11

Using the Autodesk Online Help 1-11
Using the Context Sensitive Help 1-11
Autodesk Construction Cloud 1-11
Self-Evaluation Test 1-12
Review Questions 1-12

Chapter 2: Exploring the Navigation Tools in Navisworks

Introduction 2-2
Using the Head-Up Display Feature 2-2
Setting View Orientations 2-2
The ViewCube 2-3
The Navigation Bar 2-3
Pan Tool 2-3
Zoom Tools 2-3
Orbit Tools 2-4
Look Tools 2-4
Walk and Fly Tools 2-4
SteeringWheels 2-4
Camera 2-4
Reference Views 2-5
Tutorial 1 2-5
Tutorial 2 2-16
Self-Evaluation Test 2-19
Review Questions 2-20
Exercise 1 2-21

Chapter 3: Selecting, Controlling and Reviewing Objects

Introduction 3-2
Object Selection Methods 3-2
Direct Selection Method 3-2
The Selection Tree Window 3-2
Setting Selection Resolution 3-2
Find Items Window 3-2
Quick Find 3-2
Saving Selections 3-2
Selection Sets 3-2
Search Sets 3-2
The Selection Inspector Window 3-3
Object Properties 3-3
Properties Window 3-3
Linking External Databases To Objects 3-3
Adding Database Link Locally 3-3
Adding Database Link Globally 3-3
Measure Tools 3-3
Direction Constraints for Measurements 3-4
Markup Tools 3-4

Adding Text to Markup 3-4
Drawing Cloud 3-4
Tags and Comments 3-4
Adding Tags 3-4
The Comments Window 3-4
The Find Comments Window 3-4
Comparing Models 3-4
Find Differences In 3-5
Results 3-5
Controlling the Visibility of Objects 3-5
Controlling the Appearance of the Model in the Scene View 3-5
Controlling Object Attributes 3-5
Controlling Object Attributes using Gizmo 3-5
Controlling Object Attributes Numerically 3-5
Links 3-5
Types of Links 3-5
Customizing Appearance of Links in the Scene View 3-6
Adding Links 3-6
Managing Links 3-6
The Appearance Profiler Window 3-6
Selector Area 3-6
Appearance Area 3-6
Selector List Area 3-6
Tutorial 1 3-7
Tutorial 2 3-14
Tutorial 3 3-18
Self-Evaluation Test 3-22
Review Questions 3-22
Exercise 1 3-23

Chapter 4: Viewpoints, Sections And Animations

Introduction 4-2
Working with Viewpoints 4-2
Saving Viewpoints 4-2
Organizing Viewpoints 4-2
Editing Viewpoints 4-2
Sharing Viewpoints 4-2
Sectioning in Navisworks 4-2
Creating Section Views Using Planes Mode 4-2
Transforming Section Planes 4-3
Linking Section Planes 4-3
Creating Section Views Using the Box Mode 4-3
Transforming Section Box 4-2
Animation in Navisworks 4-2
Creating Viewpoint Animations By Recording 4-3
Playing Viewpoint Animations 4-3
Combining Multiple Animations 4-4

Editing Viewpoint Animations 4-4
Sharing Animations 4-4
Tutorial 1 4-5
Tutorial 2 4-10
Tutorial 3 4-16
Self-Evaluation Test 4-23
Review Questions 4-23
Exercise 1 4-24
Exercise 2 4-25

Chapter 5: Time Liner

Introduction 5-2
Introduction to Timeliner 5-2
The Timeliner Window 5-2
Tasks Tab 5-3
Data Sources Tab 5-3
Configure Tab 5-3
Simulate Tab 5-3
Playing Simulation 5-4
Exporting Animation from the TimeLiner 5-4
Adding Animation to the TimeLiner 5-4
Adding Animation to the Entire Schedule 5-4
Adding Animation to Tasks 5-4
TimeLiner Costs 5-4
Importing Schedules From External Scheduling Software 5-4
Tutorial 1 5-5
Tutorial 2 5-16
Self-Evaluation Test 5-20
Review Questions 5-20
Exercise 1 5-21

Chapter 6: Working with Animater and Scripter

Introduction 6-2
The Animator Window 6-2
Animator Toolbar 6-2
Animator Tree View 6-3
Animator Timeline View 6-3
Creating Object Animations 6-3
Creating Object Animation Using the Animation Set 6-3
Creating Animation Using Camera 6-3
Creating Animation Using the Section Plane 6-4
Editing Keyframes 6-4
The Scripter Window 6-4
The Scripts Area 6-4
The Events Area 6-4
The Actions Area 6-5

The Properties Area 6-5
Enabling Scripts 6-5
Tutorial 1 6-6
Tutorial 2 6-16
Self-Evaluation Test 6-27
Review Questions 6-28
Exercise 1 6-29

Chapter 7: Quantification

Introduction 7-2
Quantification Workbook Window 7-2
Creating a Project 7-2
Item Catalog 7-2
Resource Catalog 7-3
Takeoff Methods 7-3
Model Takeoff 7-4
Virtual Takeoff 7-4
2D Takeoff 7-4
Tutorial 1 7-4
Tutorial 2 7-16
Self-Evaluation Test 7-24
Review Questions 7-25
Exercise 1 7-26
Exercise 2 7-27

Chapter 8: Clash Detection

Introduction 8-2
Clash Detective Window 8-2
Test Panel 8-2
Rules Tab 8-3
Select Tab 8-3
Results Tab 8-3
Report Tab 8-3
Managing Clash Tests 8-4
Merging Clash Tests 8-4
Creating Custom Clash Tests 8-4
Linking Clash Test with Timeliner and Object Animation 8-4
Time-Based Clash Test 8-4
Soft Clashing 8-4
Time-Based Soft Clashing 8-4
Customizing Clash Detective Window 8-5
Creating Clash Report 8-5
Tutorial 1 8-6
Tutorial 2 8-15
Self-Evaluation Test 8-22

Review Questions 8-23
Exercise 1 8-24

Chapter 9: Autodesk Rendering in Naviswork

Introduction 9-2
The Autodesk Rendering Window 9-2
The Rendering Toolbar 9-3
Materials Tab 9-3
Material Mapping Tab 9-3
Lighting Tab 9-3
Environments Tab 9-3
Settings Tab 9-3
Rendering using Autodesk Graphics 9-4
Rendering in Cloud 9-4
Tutorial 1 9-4
Self-Evaluation Test 9-15
Review Questions 9-16
Exercise 1 9-17

Preface

Navisworks 2024 for Novices

Navisworks 2024 for Novices: Learn by Doing textbook that has been written to cater to the needs of the students and professionals. The chapters in this textbook are structured in a pedagogical sequence, which makes the learning process very simple and effective for both the novice as well as the advanced users of Autodesk Navisworks. In this textbook, the author emphasizes on creating 4D simulation, performing clash detection, performing quantity takeoff, rendering, creating animation, and reviewing models through tutorials and exercises. In addition, the chapters have been punctuated with tips and notes, wherever necessary, to make the concepts clear, thereby enabling you to create your own innovative projects.

The concepts in this book are explained with the help of suitable examples to facilitate better understanding. The simple and lucid language used in this textbook makes it a ready reference for both the beginners and the intermediate users. In addition, the case study added at the end would immensely help readers as it gives insights into a real world BIM project.

The main features of this textbook are as follows:

- **Project-based Approach**
  The author has adopted a project-based approach and the learn-by-doing theme throughout the textbook. This approach guides the users through the process of creating the designs given in the tutorials.

- **Real-World Designs as Projects**
  The author has used real-world building designs and architectural examples as projects in this textbook so that the users can correlate them to the real-time designs.

- **Tips and Notes**
  Additional information related to various topics is provided to the users in the form of tips and notes.

- **Learning Objectives**
  The first page of every chapter summarizes the topics that are covered in that chapter.

Symbols Used in the Book

Note

The author has provided additional information to the users about the topic being discussed in the form of notes.

Tip

Special information and techniques are provided in the form of tips that help in increasing the efficiency of the users.

Formatting Conventions Used in the Textbook

Please refer to the following list for the formatting conventions used in this textbook.

- Names of tools, buttons, options, browser, palette, panels, and tabs are written in boldface.
  Example: The **Beam** tool, the **Modify** button, the **Structure** panel, the **Home** tab, the **Properties** palette, **Project Browser**, and so on.
- Names of dialog boxes, drop-downs, drop-down lists, list boxes, areas, edit boxes, check boxes, and radio buttons are written in boldface.
  Example: The **Options** dialog box, the **Column** drop-down in the **Structure** panel of the **Structure** tab, the **Name** edit box in the **Name** dialog box, the **Chain** check box in the **Options Bar**, and so on.
- Values entered in edit boxes are written in boldface.
  Example: Enter **4'** in the **Offset** edit box.
- Names of the files saved are italicized.
  Example: *c03_Industrial-Complex_tut2.rvt*

Ribbon: Modify | (Elements / Components) > Modify > Move
Shortcut Keys: MV

- When you select an element or a component, a contextual tab is displayed depending upon the entity selected. In this textbook, this contextual tab is referred to as **Modify | (Elements / Components)**

Naming Conventions Used in the Textbook

Tool

If you click on an item in a panel of the ribbon and a command is invoked to create/edit an object or perform some action, then that item is termed as tool. For example: **Select** tool and **Zoom** tool.

Button

The item in a dialog box that has a 3d shape like a button is termed as **button**. For example, **OK** button, **Cancel** button, **Apply** button, and so on.

Drop-down

A drop-down is the one in which a set of common tools are grouped together. You can identify a drop-down with a down arrow on it. These drop-downs are given a name based on the tools grouped in them. For example, **Select All** drop-down, **Windows** drop-down, and so on, refer to Figures 1 and 2.

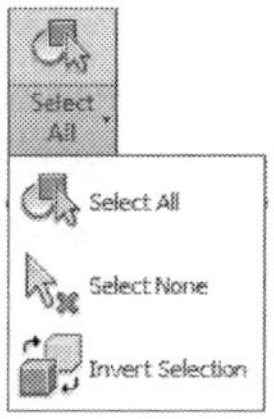

***Figure 1** The **Select All** drop-down*

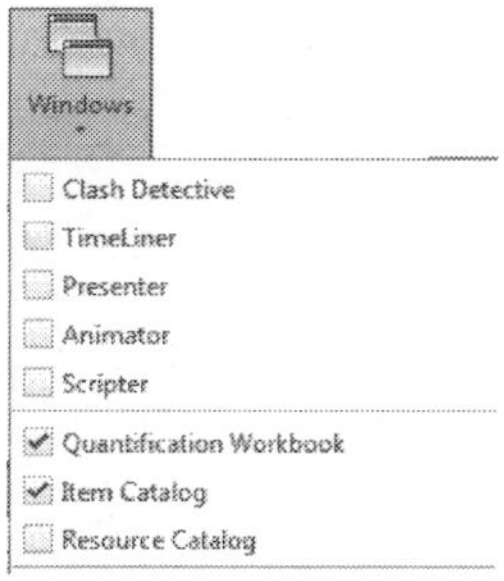

***Figure 2** The **Windows** drop-down*

Drop-down List

A drop-down list is the one in which a set of options are grouped together. You can set the parameters using these options. You can identify a drop-down list with a down arrow on it. For example, **Model Takeoff** drop-down list, **Linear Units** drop-down list, and so on, refer to Figure 3.

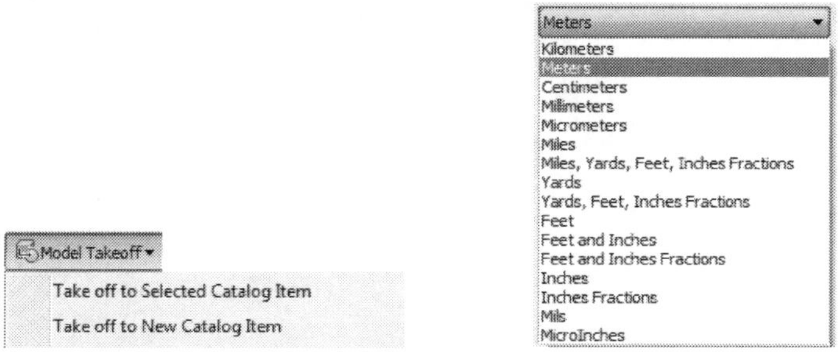

***Figure 3** The **Model Takeoff** and **Linear Units** drop-down lists*

Options

Options are the items that are available in shortcut menus, dialog boxes, drop-down lists, and so on. For example, choose the **Test Action** option from the shortcut menu displayed on right-clicking in the drawing area.

This page is intentionally left blank

Chapter 1

Introduction to Autodesk Navisworks 2024

Learning Objectives

After completing this chapter, you will be able to:

- *Understand the basic features of Autodesk Navisworks 2024*
- *Start Autodesk Navisworks 2024 software*
- *Use different components of the User Interface screen of Navisworks 2024*
- *Understand the Local and Global options in Navisworks*
- *Understand different file types in Navisworks*
- *Manage files in Navisworks*
- *Share data in Navisworks*
- *Work in different workspaces in Navisworks*
- *Access the Navisworks 2024 Help*

INTRODUCTION

Autodesk Navisworks is designed to provide an integrated environment to coordinate, share, and manage project files between multiple teams. It helps them to view the whole project and make better design conclusions, accurate construction documents, and to anticipate performance of the design. It is a powerful software which takes construction and design to new levels. It holds a building information modelling project together. The use of Autodesk Navisworks 2024 provides a competitive advantage and a higher profitability to building industry professionals.

Autodesk Navisworks has three product verticals: Navisworks Freedom, Navisworks Simulate, and Navisworks Manage.

Navisworks allows you to open the Autodesk Revit files in their native format. It also supports other file formats from other software such as CATIA, SOLIDWORKS, Autodesk ReCap, Autodesk Inventor, Faro 4.8, and so on.

In this book, you will learn about all the basic and advanced tools in Navisworks Manage.

BASIC FEATURES OF Navisworks Manage

Navisworks Manage is a complete project software which allows analysis, reviewing, coordination, and simulation. The basic features of Navisworks are listed below.

1) TimeLiner
2) Quantification
3) Clash Detection
4) Autodesk Rendering
5) Animator and Scripter

BENEFITS OF NAVISWORKS

There are various advantages of using Navisworks. Some of them are listed below:

1) Navisworks supports more than 60 different third-party application file formats.
2) Navisworks allows real-time navigation, clash detection, tolerance checking, design reviews, and project coordination with team members.
3) It improves productivity and quality by pre-visualizing the construction projects.
4) It allows rendering that helps in verifying the materials and textures that are appropriate for a proposed design.
5) It has the ability to import data from other applications to produce a single unified model.

STARTING Autodesk Navisworks 2024

You can start Navisworks by double-clicking on the Navisworks Manage 2024 icon on the desktop. To start it from the taskbar, choose **Start > All apps > Autodesk Navisworks Manage 2024 > Navisworks Manage 2024** (for Windows 11), refer to Figure 1-1.

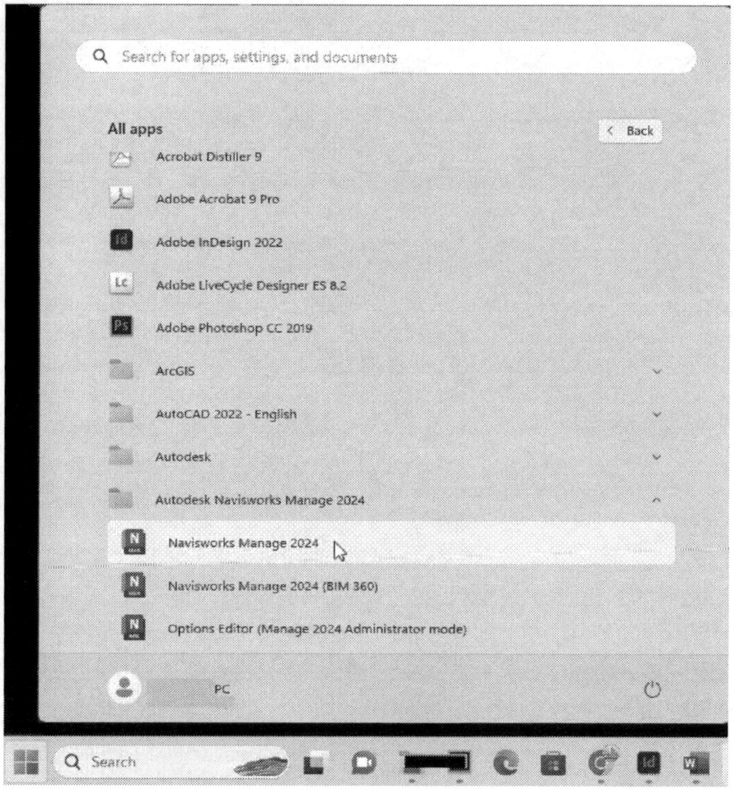

Figure 1-1 *Starting Autodesk Navisworks Manage from the taskbar*

USER INTERFACE

The user interface of Autodesk Navisworks comprises several elements such as **Application Button**, **Ribbon**, **Quick Access Toolbar**, **Info Centre**, **Scene View**, **Navigation Bar**, **Dockable Windows**, and **Status Bar**, as shown in Figure 1-2. The interface elements are exclusively designed to provide an easy access to the tools and windows.

Title Bar

The Title bar is placed at the top of the user interface. It displays the program's logo and name of the current project. It also contains the **Quick Access Toolbar** and **InfoCenter Toolbar**.

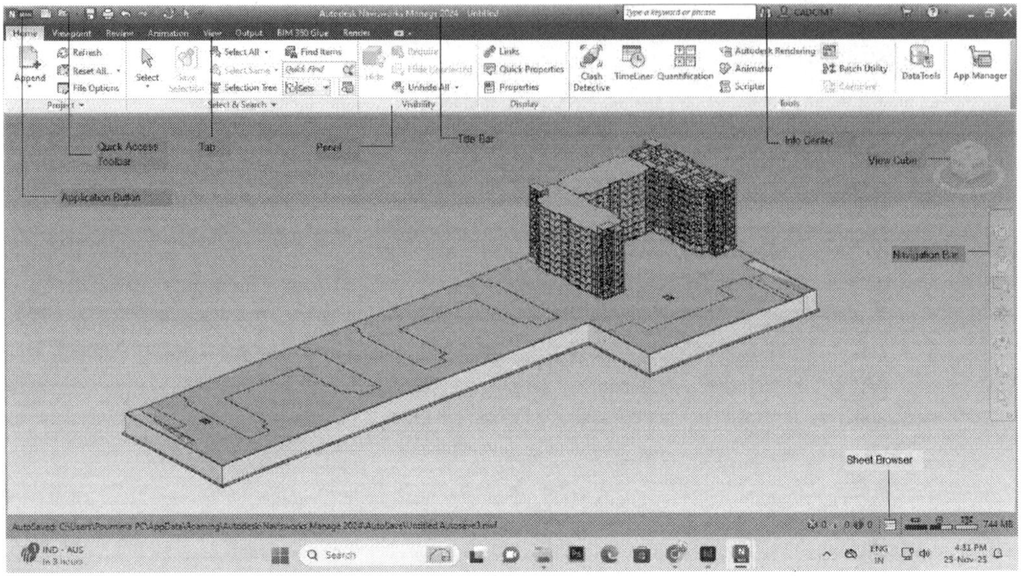

***Figure 1-2** The user interface of Autodesk Navisworks Manage 2024*

Quick Access Toolbar

The **Quick Access Toolbar** comprises several frequently used tools such as **New**, **Open**, **Save**, **Print**, **Undo**, **Redo**, **Refresh**, and **Select**. Figure 1-3 shows the **Quick Access Toolbar**.

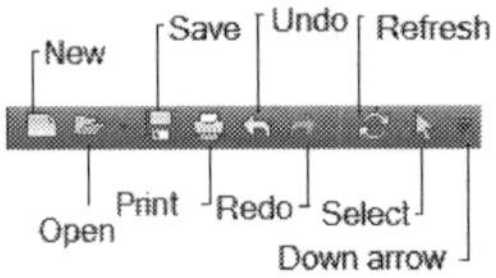

Figure 1-3** The **Quick Access Toolbar

InfoCenter

The **InfoCenter Toolbar** contains several options, which are used to access product related information. The **InfoCenter Toolbar** is aslo called as **Help Toolbar**.Figure 1-4 shows the tools in InfoCenter.

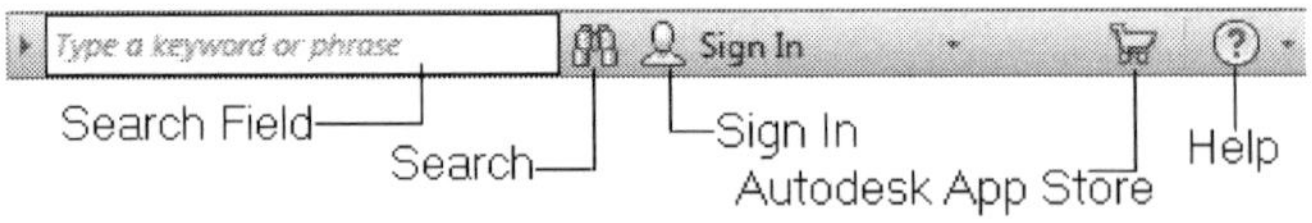

***Figure 1-4** The **InfoCenter** toolbar*

Application Menu

The Application Menu contains common tools such as **Open**, **Save**, **New**, **Save As**, **Export**, **Publish**, **Print**. To access these tools, choose the **Application Button**; the Application Menu will be displayed, as shown in Figure 1-5.

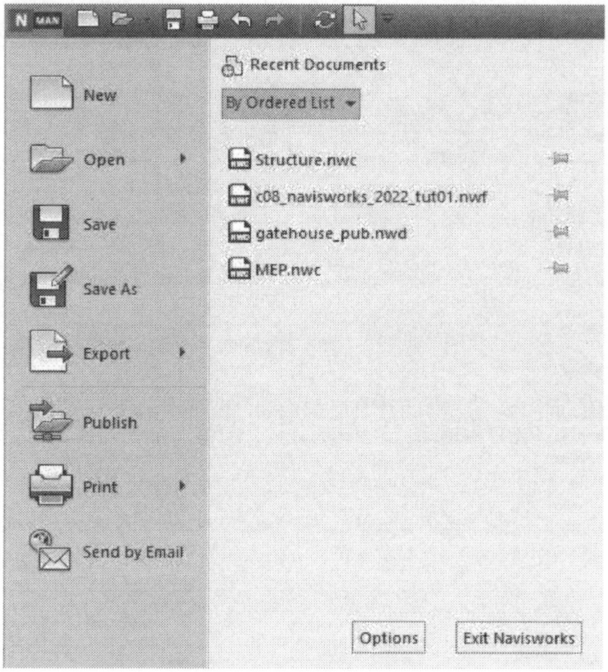

Figure 1-5 *The Application Menu*

Ribbon

The **Ribbon** is a palette in the interface that is displayed at the top of the screen, as shown in Figure 1-6. It comprises several tools and options necessary for creating a project. These tools and options are grouped together in a task-based series of panels within the tabs. These tabs and panels can be customized by the users. This can be done by moving the tabs and panels to different places in the **Ribbon**.

Figure 1-6 *Partial view of the* ***Ribbon***

Scene View

This is the area where you can actually view and work with 3D models. By default, there is only one Scene View but Navisworks allows you to split the Scene View into multiple viewports by using the options available in the **Scene View** panel, as shown in Figure 1-7. Multiple views will help you to compare the views while rendering and creating animations. drop-down.

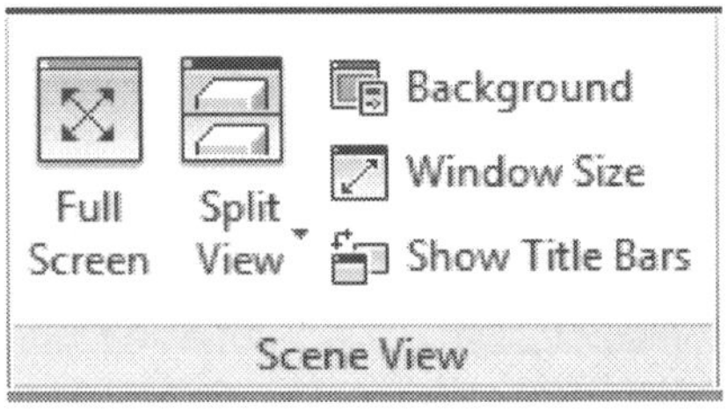

*Figure 1-7 The **Scene View** panel*

Navigation Bar

The **Navigation Bar** provides access to the tools used for navigation and orientation in a model. It contains **Steering Wheels**, **Pan**, **Zoom**, **Orbit**, **Look**, **Walk**, and **Fly** tools, as shown in Figure 1-8.

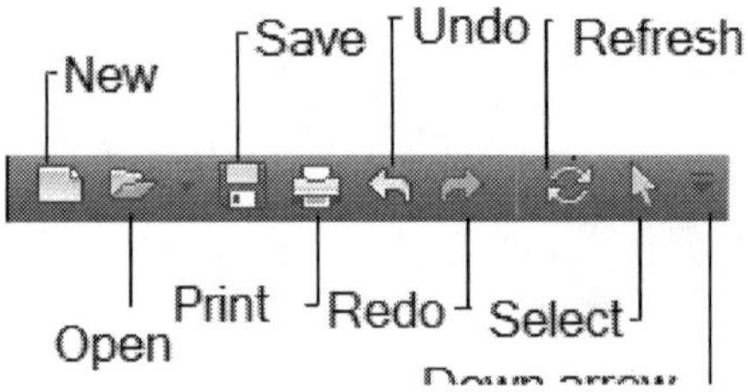

*Figure 1-8 The **Navigation Bar***

ViewCube

The ViewCube is a navigation control which is used to set the orientation of a model. It is always displayed at the top right corner in the Scene View. It comprises a cube and a compass ring at the base, as shown in Figure 1-9.

Figure 1-9 The ViewCube

Dockable Windows

Navisworks allows you to acces most of its features from the dockable windows. These dockable windows can be moved, docked, and resized in the Scene View.

Status Bar

The Status Bar is located at the bottom of the interface screen, as shown in Figure 1-10.

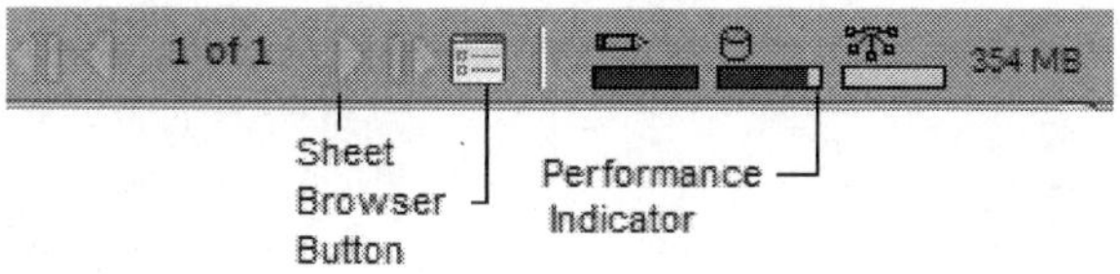

Figure 1-10 *The Status Bar*

Autodesk Navisworks DIALOG BOXES

Certain options and commands when invoked, display a dialog box. Figure 1-11 shows an example of a dialog box.

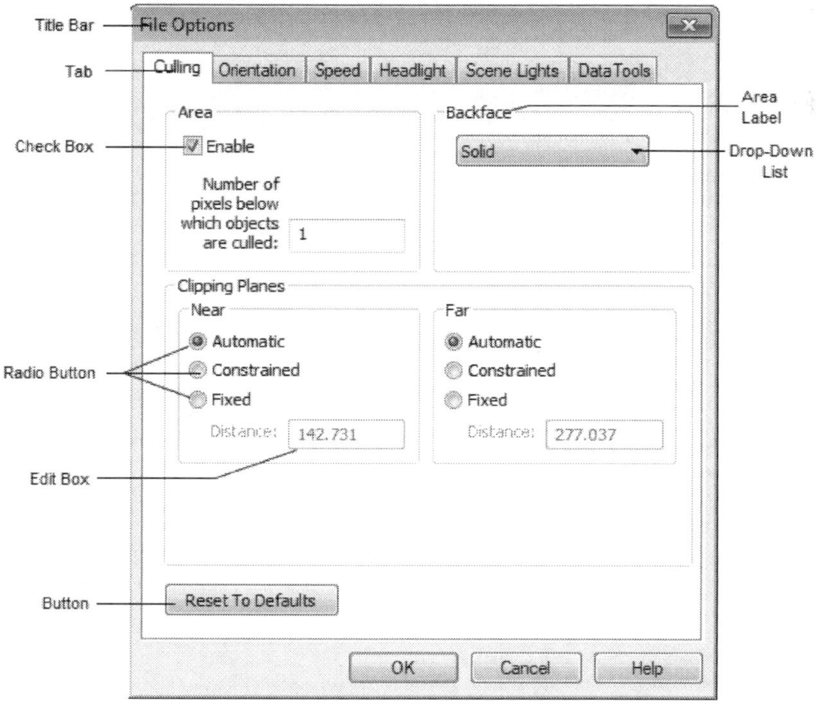

Figure 1-11 *Components of a dialog box*

CONFIGURING SETTINGS IN Navisworks

In Navisworks, you can configure the settings locally or globally by using the **File Options** dialog box and the **Options Editor** dialog box.

Configuring the Local Settings

In Navisworks, you can save the model appearance, navigation speed, and any other viewing options with different file types such as NWD and NWF. Whenever a change is made in a file type, it will be saved and reloaded when you open the file. This can be done by using various file options available in the **File Options** dialog box, refer to Figure 1-12.

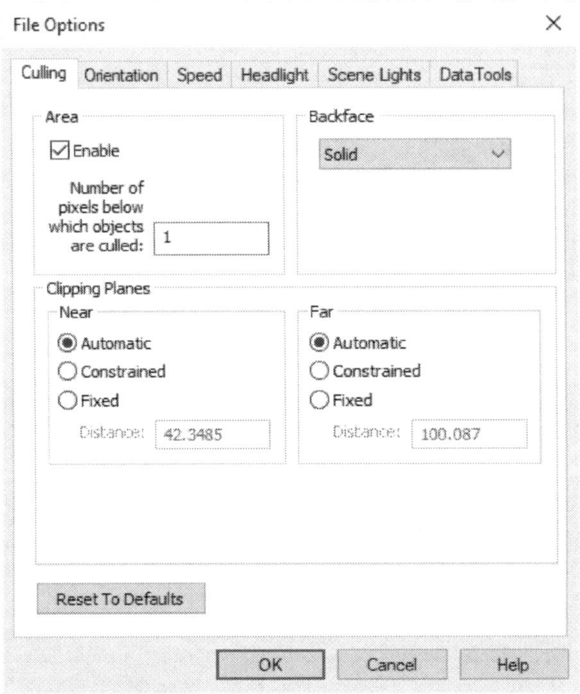

***Figure 1-12** The **File Options** dialog box*

Configuring the Global Settings

Global options can be configured and saved for all the Autodesk Navisworks sessions. You can configure these options in the **Options Editor** dialog box.

KEYBOARD SHORTCUTS

In Navisworks, keyboard shortcuts have been assigned to some of the frequently used tools to invoke them. The shortcut key corresponding to a tool is displayed when you hover the cursor over it. You can also enable the keytips by pressing the ALT key from the keyboard. As a result, various shortcut keys will be displayed in the Navisworks interface. Table 1-1 shows some of the frequently used shortcut keys in Navisworks.

Table 1-1 *Various shortcut keys used in Navisworks*

Keyboard Shortcut	Description
F1	Open the Help system
F2	Rename the selected Item
CTRL+N	Create a new file
CTRL+O	Display the **Open** dialog box
CTRL+A	Display the **Append** dialog box
CTRL+0	Turn on turntable mode
CTRL+1	Activate the **Select** tool
CTRL+2	Activate the **Walk** tool
CTRL+6	Activate the **Pan** tool
CTRL+S	Save the current file

FILE TYPES IN Navisworks

In Navisworks, you can access files originated from a variety of design and engineering applications. It has the capability to share and coordinate these file types and create a single Navisworks file with a project view of the entire model. This file contains all the geometry data and the data created by several teams, and it enables you to explore the model. Navisworks converts and compresses most of the files up to 80 percent of their original size, so that sharing and working with files may become easier. Autodesk Navisworks supports three types of file formats: **NWD File Format**, **NWC File Format**, and **NWF File Format**.

FILE READERS

Autodesk Navisworks supports a variety of additional file types such as CAD file format, laser scan formats, and other engineering design applications. The file readers provide support to these file types with several types of file formats available in Navisworks. You can modify the parameters of these file formats.

MANAGING FILES IN Navisworks

In Navisworks, you can manage files in various ways. You can append geometry and data from multiple files, merge multiple files into single file, and save, publish, and email files. Various procedures and operations used while managing files are listed below.

1) Opening Files
2) Appending Files
3) Merging Files
4) Refreshing Files
5) Saving Files
6) Publishing Files
7) Emailing Files

WORKING WITH BATCH UTILITY

The **Batch Utility** allows you to append a batch of any supported format files into a single Navisworks file. You can convert multiple design files into a single NWD file. The **Batch Utility** is integrated with the **Windows Task Scheduler** option which allows you to schedule the conversions at specified times and intervals.

Creating a List of Design Files

You can create a list of all the design files used in the current model. To do so, choose the **Batch Utility** tool from the **Tools** panel in the **Home** tab; the **Navisworks Batch Utility** dialog box will be displayed.

You can append multiple design files into a single Navisworks file, Convert multiple files into Single NWD file.

SHARING DATA IN NAVISWORKS

In Navisworks, there are three applications that you can use to share data: **Print** files, **Import** files, and **Export** files. These applications are discussed next.

Printing Files

In Navisworks, you can plot the existing view of the model using the **Print** tool. Before plotting the view, you need to see how it will appear on the sheet.

Importing Files

In Navisworks, data such as current search criteria, viewpoints, search sets, PDS review data, PDS tags, and PDS display sets can be imported.

Exporting Files

You can export data into various file formats from Navisworks such as 3D DWF/ DWFx, Google Earth KML, FBX files, and so on.

Exporting Images and Animations

In Navisworks, you can export an image as JPEG or PNG file. You can also export the rendered image in different file formats.

Export Data

You can also export clash tests information, TimeLiner schedule, search set files, viewpoint files, viewpoint reports, and tag files. The tools used for exporting this type of data are available in the **Export Data** panel of the **Output** tab.

SETTING UNITS IN NAVISWORKS

When you open a file in Navisworks, the units of the model elements in the file will be same as assigned in the original CAD application. You can also customize the display units.

NAVISWORKS WORKSPACES

In Navisworks, there are four pre-configured workspaces. You can customize these workspaces for the basic and advanced users. You can also set up a workspace as required. These workspaces are discussed next.

Safe Mode

The **Safe Mode** workspace displays the interface with minimal features.

Navisworks Extended

The **Navisworks Extended** workspace displays the interface with features recommended for advanced users.

Navisworks Standard

The **Navisworks Standard** workspace displays the interface with commonly used windows. These windows will be displayed as tabs in the Scene View.

Navisworks Minimal

The **Navisworks Minimal** is the default workspace that will be loaded when you start Navisworks. In this workspace, the dockable windows are not displayed in the Scene View and therefore you get the maximum area to work in.

AUTODESK NAVISWORKS HELP SYSTEM

The **Help** feature contains complete information about how to use the Navisworks software. With the help of user assistance, you will learn to use Navisworks efficiently. You can easily find general descriptions, procedures, definition of terms, information about tools, details about the dialog boxes, and so on, with the use of Help feature.

Using the Autodesk Online Help

In Autodesk Navisworks Manage 2024, **Autodesk Online Help** has been introduced to access various help topics online. You can access the **Autodesk Online Help** for Navisworks Manage by choosing the **Help** tool from the **Infocenter**.

Using the Context Sensitive Help

If you need help regarding a selected tool or dialog box, Autodesk Navisworks Manage provides several options to access the relevant information. Many dialog boxes contain the **Help** button that can be used to view help for the options available in that dialog box.

AUTODESK CONSTRUCTION CLOUD

Autodesk Construction Cloud™ is used as a cloud-based construction management and collaboration solution developed by Autodesk. Autodesk Construction Cloud solution connects building data, project work flows, and project teams throughout the entire building life cycle, from design to operations. Autodesk Construction Cloud is built on a integrated platform and common data environment that will empower general contractors, specialty trades, designers and owners to drive better business outcomes..

SELF-EVALUATION TEST

Answer the following questions and then compare them to those given at the end of this chapter:

1. The __________ window is used to import 2D sheets and 3D models.

2. The __________ feature is used to combine multiple models together without any duplication.

3. The __________ option is used to send Navisworks files to other users via email.

4. The __________ workspace displays the interface with minimum features.

5. The __________ comprises various navigation tools.

6. The NWC files are heavier than their original CAD applications. (T/F)

7. The NWD files contain model geometry with relevant data. (T/F)

8. Using the **Open URL** option, you can open the NWD files located on a web server. (T/F)

9. The **Batch Utility** option is used to print the Navisworks files. (T/F)

10. The NWF files do not contain any link to the original files. (T/F)

REVIEW QUESTIONS

Answer the following questions:

1. Which of the following options is used to open a file in Navisworks?

   a) **Send by Email** b) **Print**
   c) **Open** d) **Export**

2. Which of the following options is used to display only tab titles in the Ribbon?

   a) **Minimize to Panel Titles** b) **Minimize to Tabs**
   c) **Cycle through All** d) **Open URL**

3. Which of the following options is used to split the Scene View horizontally?

   a) **Split Vertical** b) **Split Horizontal**
   c) **Show Title Bars** d) **Background**

4. Which of the following options is used to display the interface for advanced users?

   a) **Safe Mode** b) **Navisworks Standard**
   c) **Navisworks Extended** d) **Navisworks Minimal**

5. Which of the following options is not available in the Application Menu?

    a) **Open** b) **Open URL**
    c) **Print** d) **Load Workspace**

6. The file options are used to save the changes within a particular file. (T/F)

7. In Navisworks, you cannot customize lighting in the Scene View. (T/F)

8. The global options are saved for all Navisworks sessions. (T/F)

9. The NWC file is a Navisworks Document File type. (T/F)

10. In Navisworks, you cannot export data in different file formats. (T/F)

Answers to Self-Evaluation Test

1. **Sheet Browser**, **2.** **Merge**, **3.** **Send by Email**, **4.** **Safe Mode**, **5.** **Navigation Bar**, **6.** F, **7.** T, **8.** T, **9.** F, **10.** F

Chapter 2

Exploring the Navigation Tools in Navisworks

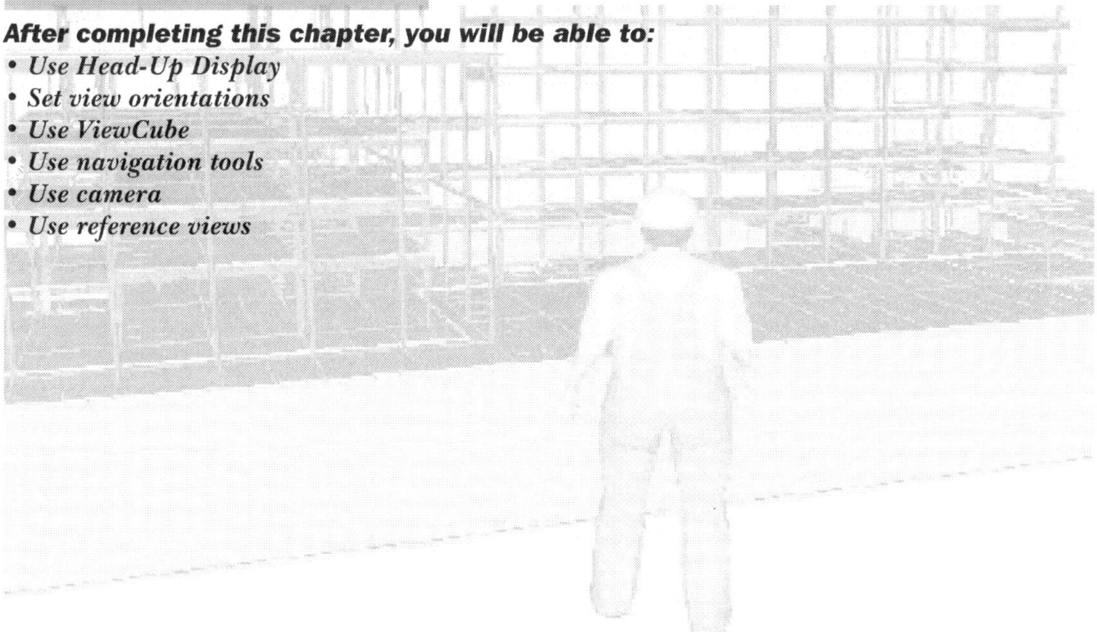

Learning Objectives

After completing this chapter, you will be able to:

- *Use Head-Up Display*
- *Set view orientations*
- *Use ViewCube*
- *Use navigation tools*
- *Use camera*
- *Use reference views*

INTRODUCTION

In Navisworks, you can explore an integrated project model using navigation tools. These tools provide a real-time navigation experience in a project. Using these navigation tools, you will be able to move around and explore the model in the Scene View. In this chapter, you will learn how to use various navigation tools and features in Navisworks.

USING THE Head-Up Display FEATURE

The **Head-Up Display** (**HUD**) feature in Navisworks helps in displaying information about the camera location and model orientation in 3D workspace. It consists of three display elements: XYZ triad, position readout, and grid position. You can display the XYZ triad by selecting the **XYZ Axes** option from the **HUD** drop-down list in the **Navigation Aids** panel of the **View** tab. On selecting this option, the information about the orientation of the model will be displayed, as shown in Figure 2-1.

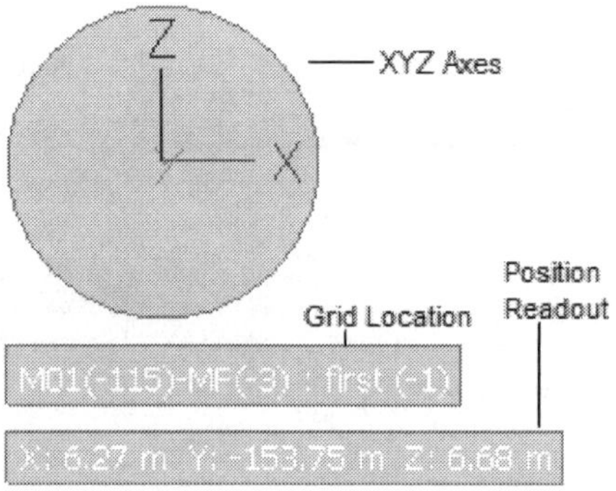

Figure 2-1 *Displaying the camera location, position, and grid location using the Head-Up Display*

SETTING VIEW ORIENTATIONS

In a project, you can set the view orientations by using various options. Although, Navisworks uses the cartesian coordinate system (X, Y, Z), the pointing direction of these axes is not fixed. By default, the Z axis points upward and the Y axis points toward north. However, you can change the up and north directions as per your requirement as refer to Figure 2-2.

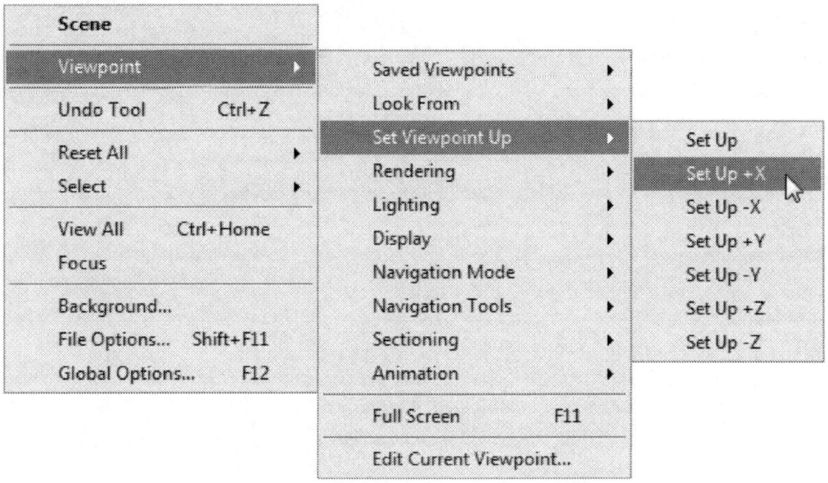

Figure 2-2 *The options to change the* ***Up*** *vectors*

THE VIEWCUBE

Navisworks provides you with a navigation control called the ViewCube that can be used to set the viewpoint of the model. You can also use the ViewCube to visually assess the current orientation of the model. The ViewCube comprises a cube, a compass ring at the base with various directions marked on it, and a home icon that helps to set the default view, as shown in Figure 2-3.

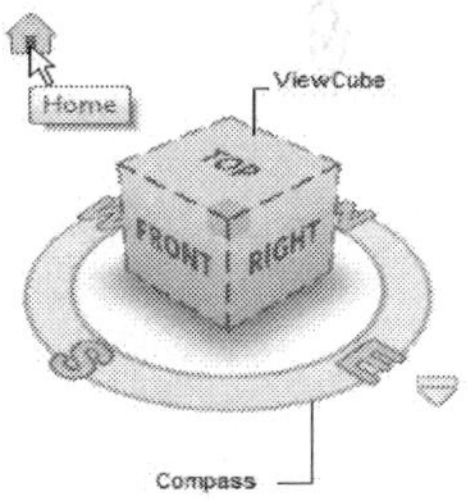

Figure 2-3 *The ViewCube*

THE NAVIGATION BAR

The Navigation Bar contains various tools such as **Pan**, **Orbit**, **Zoom**, **Look**, **Walk**, and **Fly**, refer to Figure 2-4

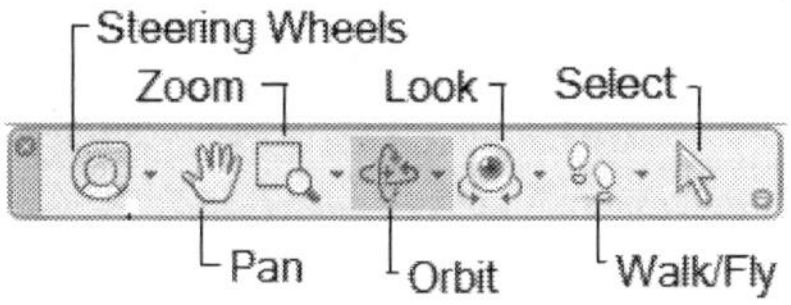

Figure 2-4 *The Navigation Bar*

Pan Tool

The **Pan** tool is used to move the model in the Scene View. During panning, the model moves along the direction of movement of the mouse.

Zoom Tools

The Zoom tools are used to increase or decrease the view size of a model. You can increase or decrease the view size of any particular item or point in a model.

Orbit Tools

The Orbit tools allow you to visually maneuver around the 3D objects to obtain different views. While orbiting, the camera location changes whereas the view remains fixed as shown in Figure 2.5.

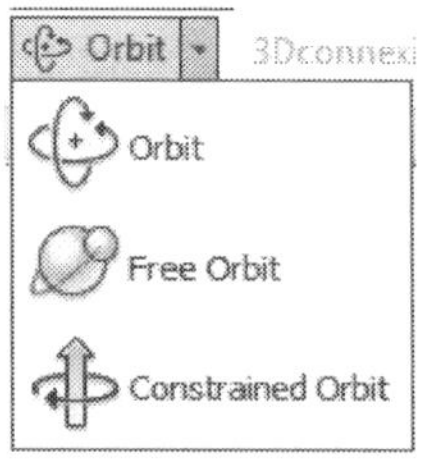

*Figure 2-5 The **Orbit** tool drop-down*

Look Tools

The Look tools are used to adjust the view of a model as shown in Figure 2-6.

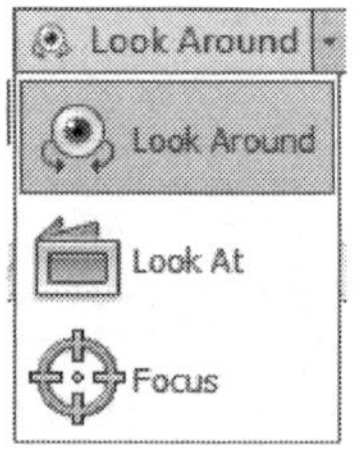

*Figure 2-6 The **Look** drop-down*

Walk and Fly Tools

The **Walk** and **Fly** tools are used to walk/fly around the model to have a real-time experience during navigation. The **Walk** tool is used to walk through a model.The **Fly** tool is used to take a birds-eye view of a model or project. This tool is used for large sites so that you can easily navigate around the outside area of your model.

SteeringWheels

The Steering Wheels are tracking menus that comprise multiple navigation tools in a single interface. These navigation tools are **Pan**, **Zoom**, **Walk**, **Look**, **Orbit**, **Rewind**, and **Up/Down**.

CAMERA

In Navisworks, you can specify various camera options such as camera projection, position and orientation during navigation. There are two types of cameras available in Navisworks: perspective and orthographic. In 3D workspace, you can use both the perspective and the orthographic cameras, but in 2D workspace, only orthographic camera can be used.

REFERENCE VIEWS

Reference views are used to acquire an overall view of your position in the whole scene. It also helps you to quickly move the camera to a different location in a model.

TUTORIALS

General instructions for downloading tutorial files:

1. Download the *c02_nws_2024_tut zip file* from *https://www.cadsofttech.com*. The path of the files follows: ***Books > Civil/Gis Books > Autodesk Navisworks > Exploring Autodesk Navisworks 2024 for Novices (Learn by Doing) > Tutorial Files.***

2. Now, save and extract the downloaded folder at the following location: *C:\ nws_2024*

Note
The default unit system used in the tutorials is metric. To change the units to imperial, select the required units from ***Options Editor > Interface > Display Units****.*

Tutorial 1 Navigating Inside the Model

In this tutorial, you will open the *c02_navisworks_2024_tutorial.nwf* file and move around the model using the **Walk**, **Collision**, **Gravity**, **Third Person**, and other navigation tools.

(Expected time: 45 min)

The following steps are required to complete this tutorial:

a. Start the Navisworks Manage session.
b. Open the *c02_navisworks_2024_tut1.nwf* file.
c. Orient the model in the Scene View using the **Orbit** and **Zoom** tools.
d. Adjust the walk speed.
e. Select Avatar and configure its dimension and position for the current viewpoint.
f. Move around in the model.
g. Move to the other floors of the building.
h. Save the project.

Starting Autodesk Navisworks 2024

1. Start **Navisworks Manage** by choosing ***Start > All apps > Autodesk Navisworks Manage 2024 > Navisworks Manage 2024*** from the taskbar; the program is loaded and the user interface screen is displayed.

Opening the Existing Model

In this section, you will open the model created in Revit software.

1. Choose the **Open** button from the Quick Access Toolbar; the **Open dialog box** is displayed.

2. In this dialog box, browse to the following location:
   C:\nws_2024\c02_nws_2024_tut

3. Select the **Navisworks File Set (*.nwf)** option from the **Files of type** drop-down list.

4. Next, select the *c02_navisworks_2024_tut1.nwf* file; the file name is displayed in the **File name** edit box.

5. Choose the **Open** button on the right of the **File name** edit box; the model is displayed in the Scene view, as shown in Figure 2-7.

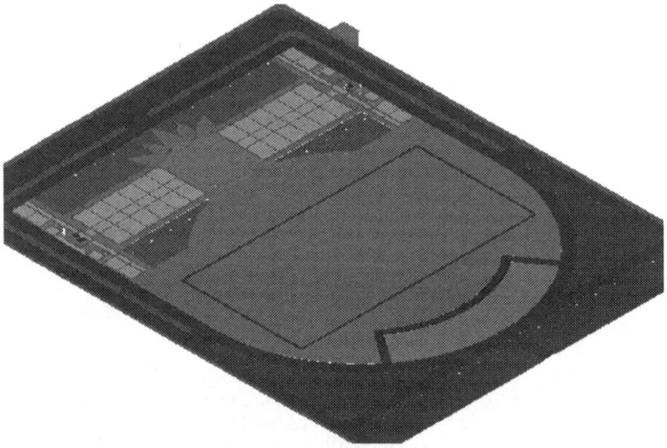

Figure 2-7 *The model opened in Navisworks*

Orienting the Model in the Scene View

In this section, you will adjust the model in the Scene View so that it fits properly in the workspace.

1. Invoke the **Orbit** tool from the **Orbit** drop-down in the **Navigate** panel of the **Viewpoint** tab; the Orbit cursor appears on the screen.

2. Press and hold the left mouse button, drag the mouse in the left direction, and then release it when the model looks similar to the one shown in Figure 2-8.

Figure 2-8 *The model after using the **Orbit** tool*

3. Next, to enlarge the viewing scale of the model, invoke the **Zoom** tool from the **Zoom** drop-down in the **Navigate** panel of the **Viewpoint** tab; the zoom cursor appears on the screen.

Note
The ***Orbit*** *and* ***Zoom*** *tools can also be invoked from the Navigation Bar.*

4. Place the zoom cursor in the model, as shown in Figure 2-9.

Figure 2-9 *Placing the* ***Zoom*** *cursor on the model*

5. Press and hold the left mouse button, drag the mouse in the forward direction, and release it when the model zooms in, as shown in Figure 2-10.

Figure 2-10 *The model after using the* ***Zoom*** *tool*

Specifying the Navigation Speed

In this section, you will adjust the walk speed while navigating inside a model.

1. Choose the **Viewpoint** tab and then expand the **Navigate** panel by clicking on the down arrow next to it.

2. Specify the walking speed **6.00** (**19ft 8.22**) in the **Linear Speed** edit box of this panel, as shown in Figure 2-11.

Figure 2-11 *Specifying the walk speed in the* ***Navigate*** *panel*

Note

You can also specify the walk speed value in the ***Linear Speed*** *edit box of the* ***Edit Current Viewpoint*** *dialog box.*

Selecting Avatars and Configuring its Parameters

In this section, you will select avatar and configure its parameters.

1. Choose the **Edit Current Viewpoint** button from the **Save, Load & Playback** panel in the **Viewpoint** tab; the **Edit Viewpoint-Current View** dialog box is displayed.

2. Choose the **Settings** button in the **Collision** area; the **Collision** dialog box is displayed, as shown in Figure 2-12.

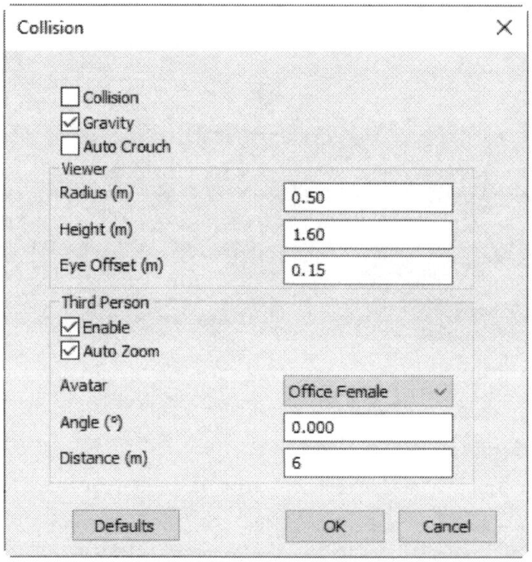

Figure 2-12 *The* ***Collision*** *dialog box*

3. In the **Collision** dialog box, select the **Enable** check box in the **Third Person** area. Ensure that the **Office Female** option is selected in the **Avatar** drop-down list.

4. Make sure that 0.00 degree is specified in the **Angle** edit box

Note that on specifying 0.00 degree value in the **Angle** edit box, the camera will be placed directly behind the avatar.

5. In the **Third Person** area, specify **6.00** (**19 ft 8.16**) in the **Distance** edit box.

6. In the **Viewer** area, enter **0.50** (**1ft 7.68**) in the **Radius** edit box.

7. Enter **1.60** (**5 ft 3.00**) in the **Height** edit box.

8. Enter **0.15** (**0 ft 6.00**) in the **Eye Offset** edit box.

The **Eye Offset** value specifies the location of camera with respect to the top of the avatar.

9. Next, choose the **OK** button; the **Collision** dialog box closes. Again choose **OK** to close the **Edit Viewpoint-Current View** dialog box.

10. Choose the **Walk** tool from the **Walk/Fly** drop-down in the **Navigate** panel of the **Viewpoint** tab; the **Walk** tool appears on the screen.

11. Press the left mouse button and click in the Scene View. Now, you can view the avatar, as shown in Figure 2-13.

Figure 2-13 *Avatar near the model*

Note

*Make sure that the **Use classic Walk** check box is cleared to disable the classic walk tool icon. This check box is available in **Options Editor > Interface > Navigation Bar**. Also, if the avatar is displayed near the roof of the South block then drag it and place it near the North block.*

12. Select the **Gravity** check box from the **Realism** drop-down in the **Navigate** panel of the **Viewpoint** tab.

13. Next, press and hold the left mouse button, drag the cursor in the forward direction, and then release it when the avatar is placed on the ground, as shown in Figure 2-14. Here, avatar faces the North block.

Note
The position of avatar is not fixed, you can navigate in and around the model as per your choice.

Figure 2-14 *Avatar standing on the ground*

Moving Around in the Model

In this section, you will walk around the model. Use the following steps to move around the model.

1. Press and hold the left mouse button, drag the mouse in the right direction until the arrow appears, refer to Figure 2-15.

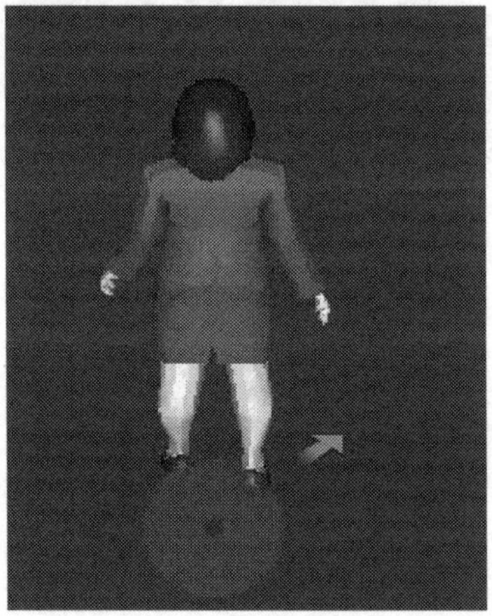

Figure 2-15 *The cursor displayed in the model*

2. Drag the cursor in the left direction and release it when the avatar faces the South block, as shown in Figure 2-16.

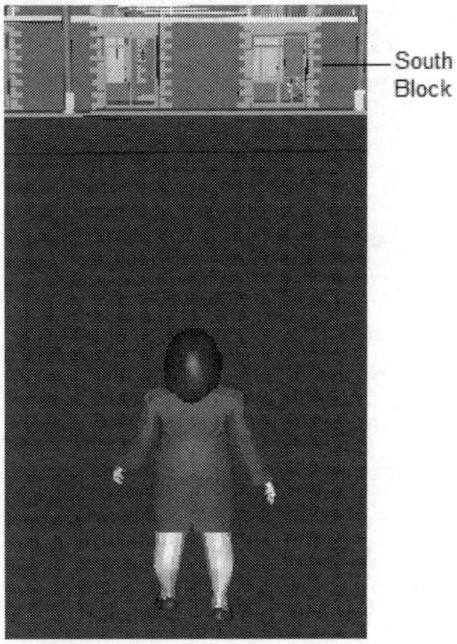

Figure 2-16 *The avatar facing the south block*

3. Drag the avatar in the forward direction and place it near the model, refer to Figure 2-17.

Figure 2-17. *Avatar moving around inside the model*

4. Drag the avatar in the left direction and place it near the fourth door from left in the model, refer to Figure 2-18.

Figure 2-18 Avatar moving in the left direction

5. Clear the **Collision** check box from the **Realism** drop-down in the **Navigate** panel of the **Viewpoint** tab.

6. Drag the avatar in the forward direction and release it when it passes through the doors, refer to Figure 2-19.

7. Drag avatar in the left direction and release it, refer to Figure 2-20.

Figure 2-19 Avatar passing through the door

Figure 2-20 Avatar moved in the left direction

8. Select the **Gravity** check box from the **Realism** drop-down in the **Navigate** panel of the **Viewpoint** tab.

9. Move the avatar toward the stairs, as shown in Figure 2-21.

10. Drag the avatar upward to climb up the stairs, as shown in Figure 2-22.

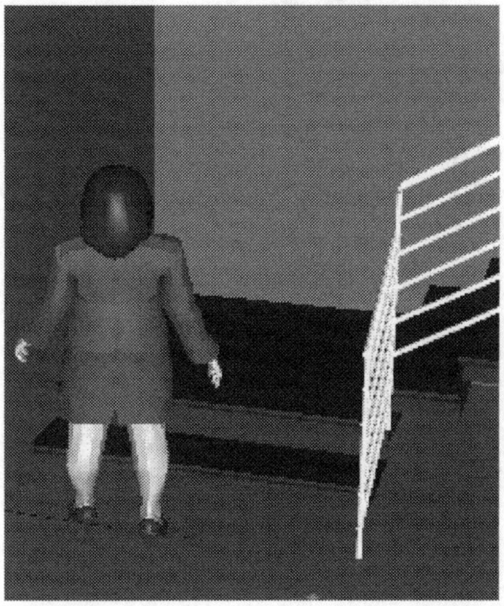

Figure 2-21 *Avatar standing near the stairs*

Figure 2-22 *Avatar walking up the stairs*

11. Drag the avatar to move up the stairs and place the avatar near the balcony door, as shown in Figure 2-23.

Figure 2-23 *Avatar standing near the balcony*

12. Clear the **Collision** check box from the **Realism** drop-down in the **Navigate** panel of the **Viewpoint** tab.

13. Next, drag the avatar to pass through the door and place it in the balcony area, as shown in Figure 2-24.

Figure 2-24 *Avatar standing in the balcony*

Navigating to the Other Levels of the Building

In this section, you will change the levels in the building using **Section View**.

1. Invoke the **Section View** window from the **Reference Views** drop-down list in the **Navigation Aids** panel of the **View** tab. The **Section View** of the building is displayed, as shown in Figure 2-25.

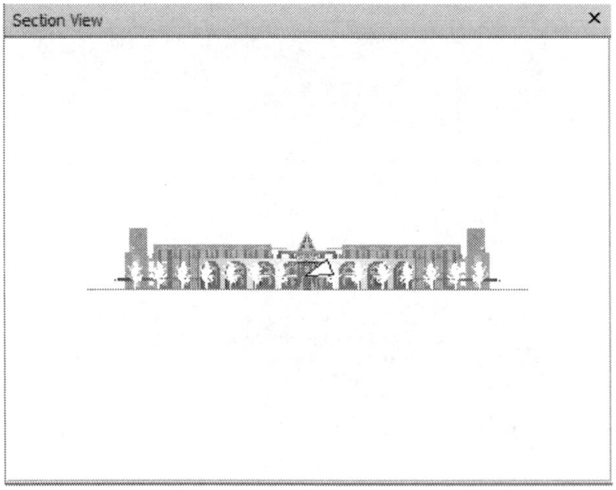

Figure 2-25 *The section view of the building in the* ***Section View*** *window*

2. In the **Section View** window, drag the triangular marker in such a way that the avatar moves to the roof of the building, as shown in Figure 2-26. Figure 2-27 shows the avatar standing on the roof. Next, close the **Section View** window.

Notice that as you move the triangular marker in the **Section View** window, the avatar moves to the roof in the Scene View.

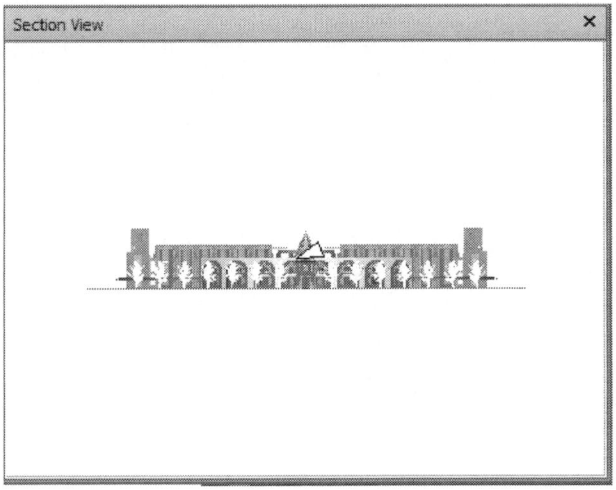

Figure 2-26 *Moving the triangular marker to the first floor of the building*

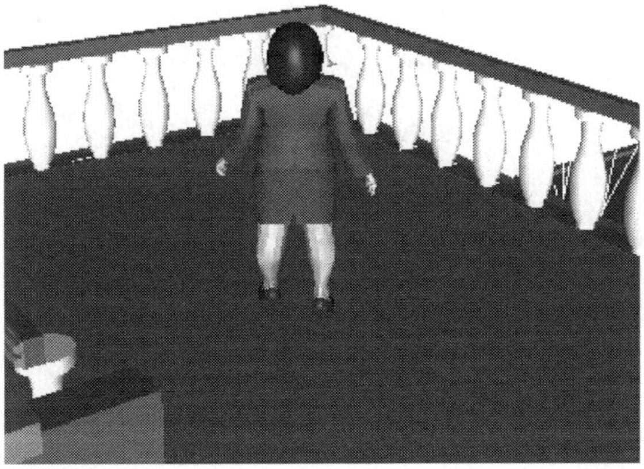

Figure 2-27 *Avatar standing on the roof of the building*

Saving the Project

In this section, you will save the project.

1. Choose **Save As** from the Application Menu; the **Save As** dialog box is displayed.

2. Browse to *nws_2024\c02_nws_2024_tut* folder and enter **c02_navisworks_2024_tut01** in the **File name** edit box. Next, choose the **Navisworks File Set** (***.nwf**) file format from the **Save as type** drop-down list, and then choose the **Save** button.

Tutorial 2 Flying Around the Model

In this tutorial, you will open the *c02_navisworks_2024_tut2.nwf* file and fly around the model using the **Fly** tool. **(Expected time: 20 min)**

The following steps are required to complete this tutorial:

a. Start the Navisworks Manage session.
b. Open the *c02_navisworks_2024_tutorial* file.
c. Adjust the model in window using the **Orbit** and **Zoom** tools.
d. Fly around the model.
e. Save the project.

Starting Autodesk Navisworks 2024

1. Navisworks Manage by choosing **Start > All Apps > Autodesk Navisworks Manage 2024 > Navisworks Manage 2024** from the taskbar

Opening the Existing File

In this section, you will open a file which is created in Revit.

1. Choose the **Open** button from the Quick Access Toolbar. Alternatively, choose **Open > Open** from the Application Menu; the **Open** file dialog box is displayed.

2. In this dialog box, browse to the following location:
   C:\nws_2024\c02_nws_2024_tut2

3. Select the **Navisworks File Set (*.nwf)** file from the **Files of type** drop-down list.

4. Next, select the *c02_navisworks_2024_tut2.nwf* file; the file name is displayed in the **File name** edit box.

5. Choose the **Open** button on the right of the **File name** edit box; the model is displayed in the Scene View, as shown in Figure 2-28.

Orienting the Model in the Scene View

In this section, you will adjust the model in the Scene View so that it fits into the Navisworks Scene View.

1. Invoke the **Orbit** tool from the **Orbit** drop-down in the **Navigate** panel of the **Viewpoint** tab; the Orbit tool cursor appears on the screen.

2. Place the cursor at the corner of the model, as shown in Figure 2-29.

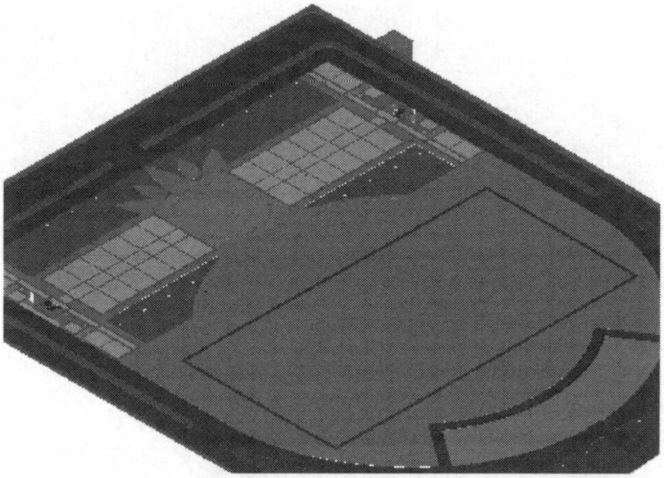

Figure 2-28 *The model displayed in the Scene View*

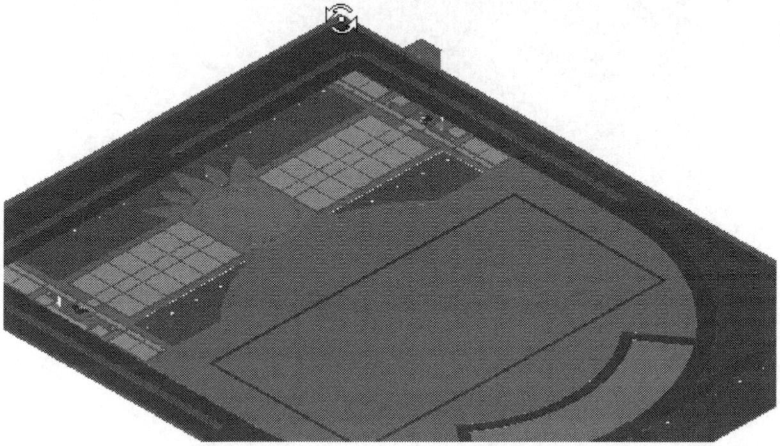

Figure 2-29 *Placing the cursor on the model*

3. Press and hold the left mouse button, drag the mouse in the downward direction, and then release it when the model is rotated, as shown in Figure 2-30.

4. Now drag the model in the left direction and release it when the model looks like, as shown in Figure 2-31.

Figure 2-30 *Model after rotating*

Figure 2-31 *The model after using the* ***Orbit*** *tool*

5. Invoke the **Zoom** tool from the **Zoom** drop-down in the **Navigate** panel of the **Viewpoint** tab; the **Zoom** tool cursor appears on the screen.

6. Press and hold the left mouse button, drag the mouse in the forward direction, and then release it when the model is zoomed, as shown in Figure 2-32.

Figure 2-32 *The model after using the* ***Zoom*** *tool*

Flying Around the Model

In this section, you will fly around the model using the **Fly** tool.

1. Invoke the **Fly** tool from the **Walk/Fly** drop-down list in the **Navigate** panel of the **Viewpoint** tab; the flight cursor appears on your screen.

2. Hold the left mouse button and drag the mouse in the right direction to fly around the model. Figure 2-33 shows a view in the fly mode.

Note

*Use the **Pan**, **Orbit**, and **Zoom** tools while flying around the model to adjust the model in such a way that you can clearly view the model and it becomes easier to fly around the model.*

Figure 2-33 *Flying around the corner of the School building*

Saving the Project

In this section, you will save the project.

1. Choose the **Save As** from the Application Menu; the **Save As** dialog box is displayed.

2. Browse to *nws_2024\c02_nws_2024_tut* folder and enter **c02_navisworks_2024_tut02** in the **File name** edit box. Next, choose the **Navisworks File Set** (***.nwf**) file format from the **Save as type** drop-down list, and then choose the **Save** button.

SELF-EVALUATION TEST

Answer the following questions and then compare them to those given at the end of this chapter:

1. The __________ tool is used to zoom a specified area.

2. You can link the position of Navigation Bar with the __________ .

3. The __________ tool is used to rotate the model freely around the pivot point in any direction.

4. The **Edit Viewpoint - Current View** dialog box can be displayed by choosing the __________ button.

5. The Section View can be displayed by selecting an option from the __________ drop-down list.

6. The docking position of the **Navigation** Bar cannot be changed. (T/F)

7. The **Orbit** tool is used to move the given model in the Scene View. (T/F)

8. The **Look At** tool is used to look around the model. (T/F)

9. In Navisworks, the walk speed cannot be changed while navigating inside a model. (T/F)

10. In Navisworks, the **Gravity** option can be used while flying around a model. (T/F)

REVIEW QUESTIONS

Answer the following questions:

1. Which of the following options is used to pass the avatar through those objects that are too low to walk under?

   a) **Gravity** b) **Crouch**
   c) **Collision** d) **Fly**

2. Which of the following options is used for invoking the third person view?

   a) **Camera** b) **Orbit**
   c) **Avatar** d) **Zoom**

3. Which of the following tools is used to change the viewing scale of the model?

   a) **Orbit** b) **Pan**
   c) **Fly** d) **Zoom**

4. Which of the following options is used to display the absolute X, Y, Z position of the camera?

   a) **XYZ Axes** b) **Grid Location**
   c) **Position Readout** d) **Section View**

5. You can specify the radius of an avatar by entering values in the **Eye Offset** boxes. (T/F)

6. Clearing the **Collision** check box will enable the avatar to pass through the doors and columns. (T/F)

7. You cannot change the height of the avatar. (T/F)

8. Section View is used to display the top view of a model. (T/F)

9. In Navisworks, the **Walk** and **Fly** tools can be used simultaneously. (T/F)

10. The **Pan** tool is used to rotate a model. (T/F)

EXERCISE

Exercise 1 Navigating Inside the Residence Building

Download and open the *c02_navisworks_2024_ex1.nwf* file from *https://www.cadsofttech.com*. Navigate inside the Residence Building model, shown in Figure 2-34, by using various navigation tools. **(Expected time: 30 min)**

The steps required to complete this exercise are given next:

1. Open the *c02_navisworks_2024_ex1* file.
2. Orient the model using the **Orbit** and **Zoom** tool.
3. Adjust the walking speed.
4. Select avatar and configure its dimension and position for the current viewpoint.
5. Navigate inside the model and on the other floors of the building.
6. Save the file with the name *c02_navisworks_2024_ex01*.

Figure 2-34 *The Residence Building*

Answers to Self-Evaluation Test

1. **Zoom Window**, **2.** **ViewCube**, **3.** **Free Orbit**, **4.** **Edit Current Viewpoint**, **5.** **Reference Views**, **6.** F, **7.** F, **8.** F, **9.** F, **10.** F

Chapter 3

Selecting, Controlling, and Reviewing Objects

Learning Objectives

After completing this chapter, you will be able to:

- *Use direct selection tools*
- *Use different types of windows for selecting objects*
- *Save selected objects*
- *Analyze object properties*
- *Link databases to objects*
- *Control visibility of objects*
- *Control the appearance of model*
- *Control object attributes*
- *Use the Measure and Markup tools*
- *Add tags and comments*
- *Use links*
- *Use the Appearance Profiler window*

INTRODUCTION

In the previous chapter, you learned to use various navigation tools in a project. In this chapter, you will learn various methods and tools to select, control, and review objects.

OBJECT SELECTION METHODS

In order to work on an object or a group of objects, you need to select them. In Navisworks, you can do so by using the Direct Selection method. You can also select an object or group of objects by using the **Selection Tree** window and the **Find Items** window.

Direct Selection Method

In the Direct Selection method, you can select an object or group of objects directly from the Scene View by using the direct selection tools. The direct selection tools are available in the **Select & Search** panel of the **Home** tab.

The Selection Tree Window

In Navisworks, you can select the object(s) from the **Selection Tree** window. To use this window, choose the **Selection Tree** button from the **Select & Search** panel in the **Home** tab.

Setting Selection Resolution

In Navisworks, at times, when you select an object in the Scene View, then either the entire model or a particular layer will be selected but not the single entity.

Find Items Window

In a project, you may need to select or search an object on the basis of its property.

Quick Find

The **Quick Find** text box is used to quickly find and select objects in a model. This text box is available in the **Select & Search** panel of the **Home** tab.

SAVING SELECTIONS

After selecting an object or a group of objects, you can save them as sets. They can be saved either by creating selection sets or by creating search sets.

Selection Sets

The selection sets are a group of objects selected either from the Scene View or from the **Selection Tree** window.

Search Sets

The search sets are a group of searched objects. When you run a search in the **Find Items** window, the searched objects will be selected and highlighted in blue color in the Scene View as well as in the **Selection Tree** window.

The Selection Inspector Window

In a project, when you select an object or a group of objects, they will be displayed in the **Selection Inspector** window along with their properties.

OBJECT PROPERTIES

In a project, you can analyze the properties of the selected object by using the **Properties** window. You can also customize the properties of selected object in the **Properties** window.

Properties Window

After selecting an object in the Scene View, you need to invoke the **Properties** window for analyzing the object properties.

LINKING EXTERNAL DATABASES TO OBJECTS

While working on a project, you may need to add some extra properties to the model object. These properties are displayed in the **Properties** window.

Adding Database Link Locally

You can add external databases locally by using the **File Options** dialog box.

Adding Database Link Globally

To add a database link globally, choose the **DataTools** tool from the **Tools** panel of the **Home** tab; the **DataTools** dialog box will be displayed.

MEASURE TOOLS

In a project, you can perform several functions such as calculating distances, angle, and area. To perform these functions, you can use measure tools. Before measuring an object, you need to adjust the cursor snapping. To do so, invoke the **Options Editor** dialog box by choosing the **Options** button from the **Application Menu**. Expand the **Interface** node in the left pane of this dialog box and select the **Snapping** option; several options will be displayed in the right pane of the dialog box. Measure tools are listed below :

1) Point to Point Tool
2) Point to Multiple Points Tool
3) Point Line Tool
4) Accumulate Tool
5) Angle Tool
6) Area Tool
7) Shortest Distance Tool
8) Clear Tool
9) Convert to Markup Tool

DIRECTION CONSTRAINTS FOR MEASUREMENTS

In a project, you can constrain the direction of measuring objects. The constraints can be applied in X, Y, and Z directions and they can also be applied in a direction parallel or perpendicular to the surface of an object.The tools used for constraining the direction are listed below:

1) X Axis Tool
2) Y Axis Tool
3) Z Axis Tool
4) Perpendicular Tool
5) Parallel Tool

MARKUP TOOLS

In a project, you can mark up the viewpoints and clash results by adding texts and drawing revision clouds. These mark ups are added by using the markup tools.

Adding Text to Markup

In a project, you can mark or annotate a saved viewpoint and clash results by adding text.

Drawing Cloud

In a project, you can mark a saved viewpoint and clash result by drawing cloud, line, ellipse, and so on.

TAGS AND COMMENTS

In Navisworks, you can tag anything in the Scene View which allows you to add comments or notes to the viewpoints. The comments added to the saved viewpoints and animations can be used for reference later

Adding Tags

In a project, you can tag anything in the Scene View.

The Comments Window

In a project, when you add comments to viewpoints or results, you can view them in the **Comments** window. In this window, you can add all the comments related to the model and can obtain any information quickly. The data will be displayed in a separate window rather than the Scene View.

The Find Comments Window

The **Find Comments** window is a dockable window. Using the options in this window, you can search for the added comments.

COMPARING MODELS

While working on a project, sometimes you may need to find the differences between two elements in a model. Alternatively, you may need to compare two models of different versions. In Navisworks, you can find out the differences or compare two models by using the **Compare** tool.

Find Differences In

In the **Find Differences In** area of the dialog box, select the check boxes corresponding to the categories in which you want to find the differences.

Results

In the **Results** area, there are various options that allow you to specify how you want to display the comparison results.

CONTROLLING THE VISIBILITY OF OBJECTS

In a project, you may need to control the visibility of some objects in the Scene View. In Navisworks, you have several options to control the visibility of objects in the model. You can hide and display an object or a group of objects in the model.

CONTROLLING THE APPEARANCE OF MODEL IN THE SCENE VIEW

You can control the appearance of a model in the Scene View. To control the display of your model in the Scene View, use the tools available in the **Render Style** panel of the **Viewpoint** tab.

CONTROLLING OBJECT ATTRIBUTES

In a project, you can control the object transformation and its appearance. All these changes are displayed in the Scene View. You can alter the object's position, size, and rotation using the visual manipulation tools (Gizmos). You can also make these changes numerically.

Controlling Object Attributes Using Gizmo

You can move, rotate, and resize the objects using gizmo.

Controlling Object Attributes Numerically

You can also move, rotate, and scale the selected objects numerically.

LINKS

Navisworks uses several sources of links such as the links converted from CAD files, links added by the users, and links which are automatically generated by the program, like viewpoint links. To display these links, choose the **Links** button from the **Display** panel in the **Home** tab; all links will be displayed as icons in the Scene View.

Types of Links

There are two types of links:

1) Standard Defined
2) User Defined

Customizing Appearance of Links in the Scene View

To avoid the confusion arising due to the large number of links displayed in the Scene View, you can customize the appearance of links displayed in the Scene View by using options in the **Options Editor** dialog box.

Adding Links

The links can be appended from various data sources such as spreadsheets, webpages, audio, and video files. You can attach multiple links to a single object.

Managing Links

In Navisworks, you can edit, reset, and delete all the links except those which are automatically generated such as links which point to the clash results, viewpoints, TimeLiner tasks.

THE APPEARANCE PROFILER WINDOW

In a project, you can customize the appearance of a model by defining the custom display settings based on the property data or selection/search sets using the **Appearance Profiler** window.

Selector Area

The left pane of the **Appearance Profiler** window is the **Selector** area. In this area, there are two tabs: **By Property** and **By Set**.

Appearance Area

The **Appearance** area is located at the right corner in the **Appearance Profiler** window.

Selector List Area

The right pane of the **Appearance Profiler** window is the **Selector** list. This area contains the list of all added selectors.

TUTORIALS

General instructions for downloading tutorial files:

1. Download the *c03_navisworks_tut* zip file for the tutorial from *https://www.cadsofttech.com*. The path of the file is as follows: ***Books > Civil/GIS Books > Autodesk Navisworks > Exploring Autodesk Navisworks 2024 for Novices (Learn By Doing) > Tutorial Files.***

2. Now, save and extract the downloaded folder at the following location:
   C:\nws_2024

Note

You can change the units from metric to imperial. To do so, select the required units from ***Options Editor > Interface > Display Units****.*

Tutorial 1 Taking Measurements

In this tutorial, you will open the file *c03_navisworks_2024_tut1.nwf.* Next, calculate the distance between mullions in the door by using various measuring tools such as the **Point to Point** tool, the **Point Line** tool, and the **Accumulate** tool. **(Expected time: 30 min)**

The following steps are required to complete this tutorial:

a. Open the file *c03_navisworks_2024_tut1.nwf.*
b. Display the saved views.
c. Measure distance using the **Point to Point** tool.
d. Measure distance using the **Point Line** tool.
e. Measure distance using the **Accumulate** tool.
f. Save the project.

Opening the File

In this section, you will open the file in Navisworks.

1. Choose the **Open** button from the Quick Access Toolbar; the **Open** dialog box is displayed.

2. In this dialog box, browse to the following location:
   C:\nws_2024\c03_nws_2024_tut.

3. Select **Navisworks File Set (*.nwf)** from the **Files of type** drop-down list.

4. Select the file *c03_navisworks_2024_tut1.nwf* from the displayed list of files; the file name is displayed in the **File name** edit box.

5. Choose the **Open** button from the dialog box; the model is displayed in the Scene View, as shown in Figure 3-1

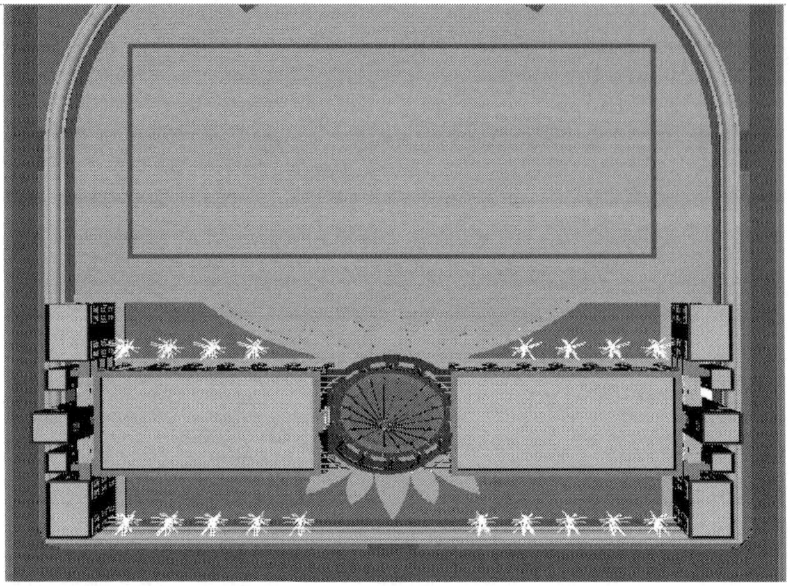

Figure 3-1 Partial view of the model opened in Navisworks

Displaying the Saved Views

In this section, you will display the saved views by using the **Saved Viewpoints** window.

1. Choose the **Saved Viewpoints Dialog Launcher** button from the **Save, Load & Playback** panel of the **Viewpoint** tab; the **Saved Viewpoints** window is displayed, as shown in Figure 3-2.

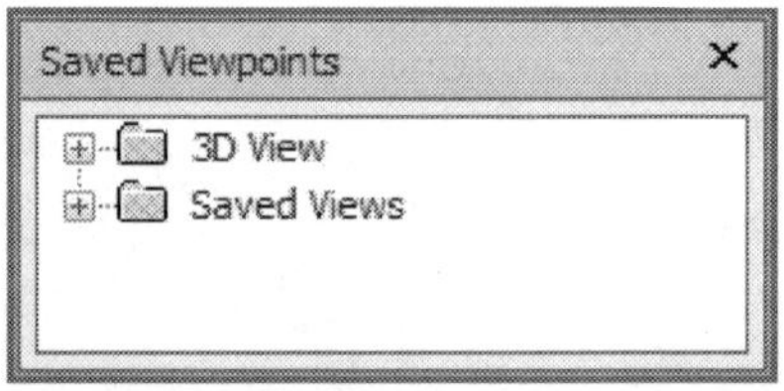

*Figure 3-2 The **Saved Viewpoints** window*

2. In the **Saved Viewpoints** window, click on the **+** icon corresponding to the **Saved Views** viewpoint folder; the folder is expanded, as shown in Figure 3-3.

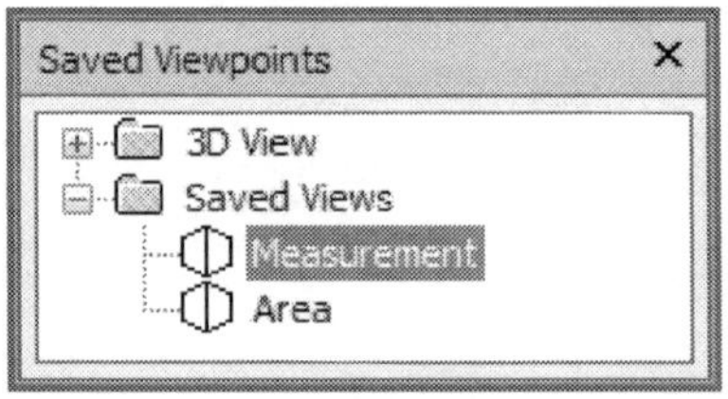

*Figure 3-3 Expanded view of the **Saved Views** folder*

3. Select the **Measurement** option from the **Saved Views** folder, refer to Figure 3-3; the view is displayed in the Scene View, as shown in Figure 3-4. Close the **Saved Viewpoints** window.

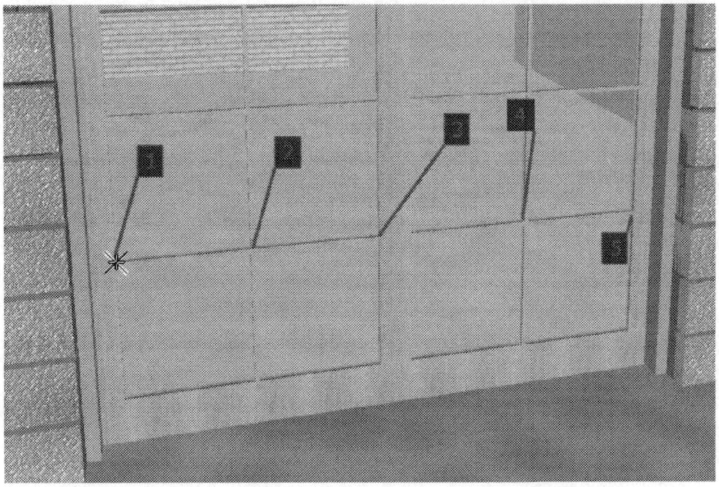

***Figure 3-4** The **Measurement** view in the Scene View*

Measuring Distance Using the Point to Point Tool

In this section, you will measure distance using the **Point to Point** tool.

1. Invoke the **Point to Point** tool from **Review > Measure > Measure** drop-down; the cursor is displayed on the screen.

2. Move and place the cursor at point **1** and click when the snap to vertex cursor appears, as shown in Figure 3-5.

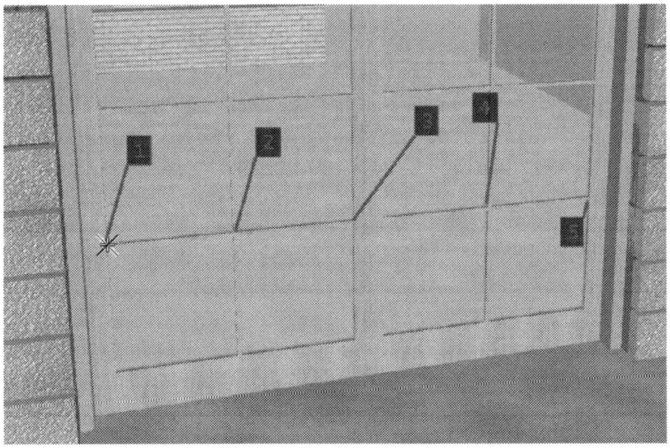

Figure 3-5** The cursor placed at point **1

3. Next, place the cursor on point **2** and click when the snap to vertex cursor appears; the calculated distance is displayed in the Scene View, as shown in Figure 3-6.]

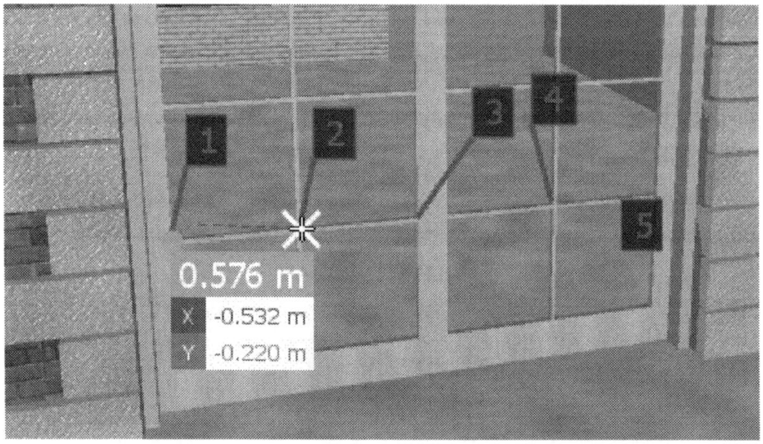

Figure 3-6 *Distance calculated between point **1** and **2***

Measuring Distances Using the Point Line Tool

In this section, you will measure distance using the **Point Line** tool.

1. Invoke the **Point Line** tool from the **Review > Measure > Measure** drop-down; the cursor appears on the screen and the previously calculated distance disappears.

2. Place the cursor on point **1** and click when the snap to vertex cursor appears, as shown in Figure 3-7.

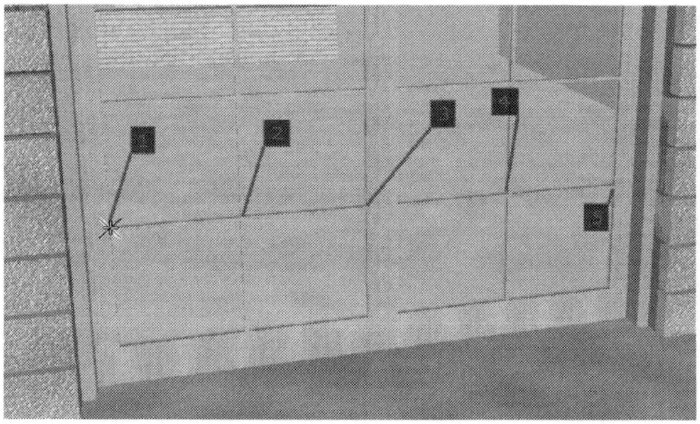

Figure 3-7 *The cursor placed at point **1***

3. Next, place the cursor on point **2** and click when the snap to vertex cursor appears; the calculated distance is displayed in the Scene View, as shown in Figure 3-8.

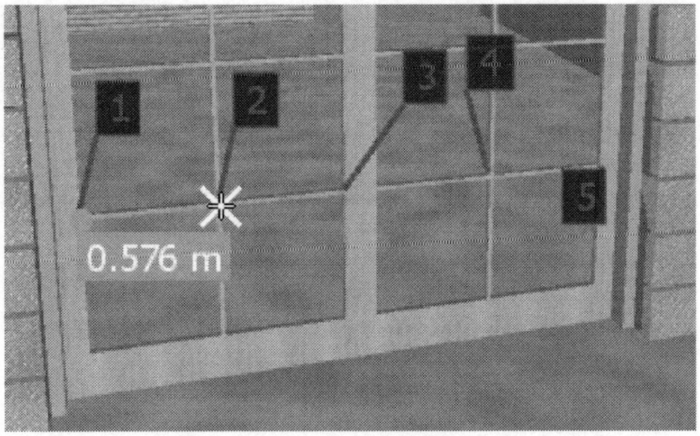

*Figure 3-8 Distance calculated between point **1** and **2***

4. Repeat the procedure followed in steps 1 through 3 to calculate the total distance between **1** and **5**; the calculated distance is displayed in the Scene View, as shown in Figure 3-9.

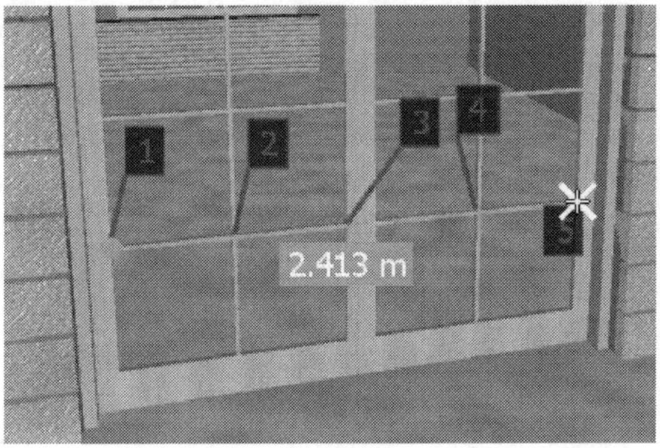

*Figure 3-9 The total distance calculated between point **1** and **5***

Measuring Distances Using the Accumulate Tool

In this section, you will measure distances using the **Accumulate** tool.

1. Invoke the **Accumulate** tool from **Review > Measure > Measure** drop-down; the cursor appears on the screen and the previously calculated distance disappears.

2. Place the cursor on point **1** and click when the snap to vertex cursor appears, as shown in Figure 3-10.

Figure 3-10 *The cursor placed at point* ***1***

3. Next, place the cursor on point **2** and click when the snap to vertex cursor appears; the calculated distance is displayed in the Scene View, as shown in Figure 3-11.

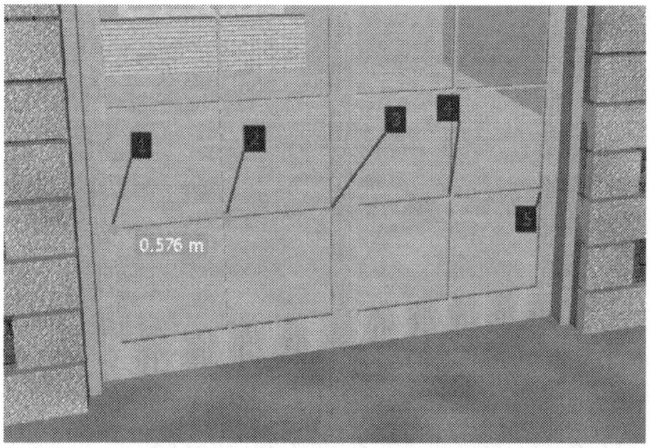

Figure 3-11 *Distance calculated between point* ***1*** *and* ***2***

4. Again, click on the point **2** and then click on point **3**; the calculated distance is displayed in the Scene View, as shown in Figure 3-12.

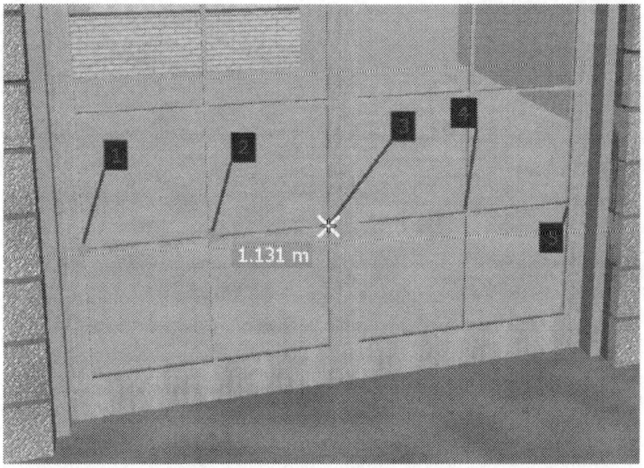

Figure 3-12 Distance calculated between point 1 and 3

5. Repeat the procedure followed in step 3 and calculate the distance between point **1** and **4,** and point **1** and **5**, refer to Figure 3-13 and Figure 3-14.

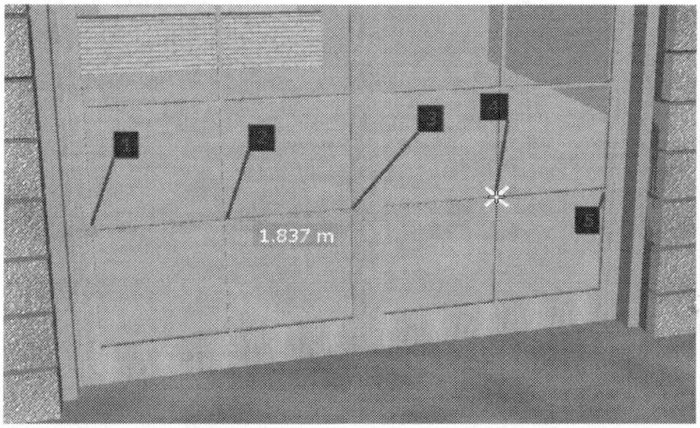

Figure 3-13 Distance calculated between point 1 and 4

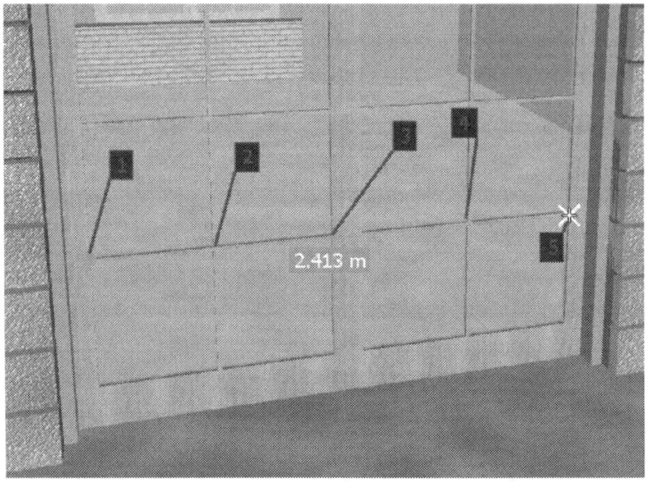

Figure 3-14 *Distance calculated between point **1** and **5***

Note

*To save the dimensions, choose the **Convert to Markup** tool from the **Measure** panel in the **Review** tab; the view will be saved in the **Saved Viewpoints** window.*

Saving the Project

In this section, you will save the project.

1. Choose **Save As** from the Application Menu; the **Save As** dialog box is displayed.

2. Browse to *Navisworks_2024* folder and enter **c03_navisworks_2024_tut01** in the **File name** edit box.

3. Next, select the **Navisworks File Set (*.nwf)** file format from the **Save as type** drop-down list and then choose the **Save** button; the project is saved.

Tutorial 2 Creating Tags and Calculating Area

In this tutorial, you will open the file *c03_navisworks_2024_tut2.nwf.* Next, you will add tags to four corners of the roof and calculate the area of roof using the **Add Tag** and **Area** tools.

(Expected time: 45 min)

The following steps are required to complete this tutorial:

a. Open the file *c03_navisworks_2024_tut2.nwf.*
b. Display the saved views.
c. Add tags.
d. Calculate area.
e. Add text.
f. Save the project.

Opening the File

In this section, you will open the file in Navisworks.

1. Choose the **Open** button from the Quick Access Toolbar; the **Open** dialog box is displayed.

2. In this dialog box, browse to the following location: *C:\nws_2024\c03_nws_2024_tut*.

3. Select **Navisworks File Set (*.nwf)** from the **Files of type** drop-down list.

4. Select the file *c03_navisworks_2024_tut2* from the displayed list of files; the file name is displayed in the **File name** edit box.

5. Choose the **Open** button from the dialog box; the model is displayed in the Scene View, refer to Figure 3-1 (same as Tutorial 1).

Displaying the Saved Views

In this section, you will display the saved views by using the **Saved Viewpoints** window.

1. Choose the **Saved Viewpoints Dialog Launcher** button from the **Save, Load & Playback** panel of the **Viewpoint** tab; the **Saved Viewpoints** window is displayed, refer to Figure 3-2.

2. In the **Saved Viewpoints** window, click on the **+** icon corresponding to **Saved Views** viewpoint folder; the folder expands, refer to Figure 3-3.

3. Select the **Area** option from the **Saved Views** folder, as shown in Figure 3-15; the view is displayed in the Scene View, as shown in Figure 3-16. Now, close the **Saved Viewpoints** window.

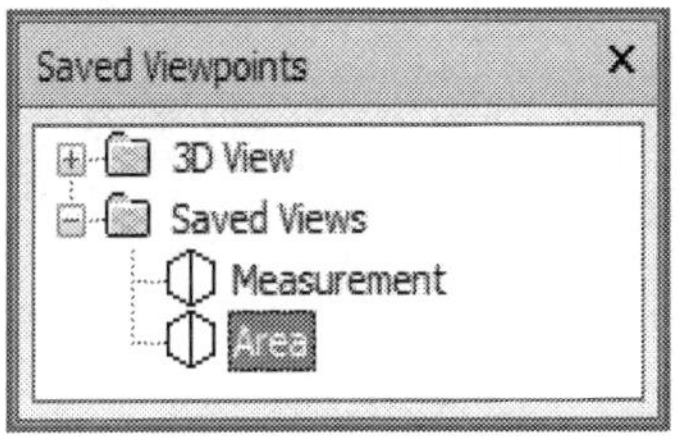

Figure 3-15 *Expanded view of the* ***Saved Viewpoints*** *folder*

***Figure 3-16** The Area view in the Scene View*

Adding Tags

In this section, you will add tags to the four corners of the roof using the **Add Tag** tool.

1. Choose the **Add Tag** tool from the **Tags** panel in the **Review** tab; the shape of the cursor is changed to pencil.

2. Place the cursor at the corner of **Wing A**, as shown in Figure 3-17.

3. Double-click on the specified location; the tag 6 is created and the **Add Comment** dialog box is displayed.

4. In the dialog box, add the comment **Corner point 1** in the text box and choose the **OK** button; the dialog box is closed and the tag is created, as shown in Figure 3-18.

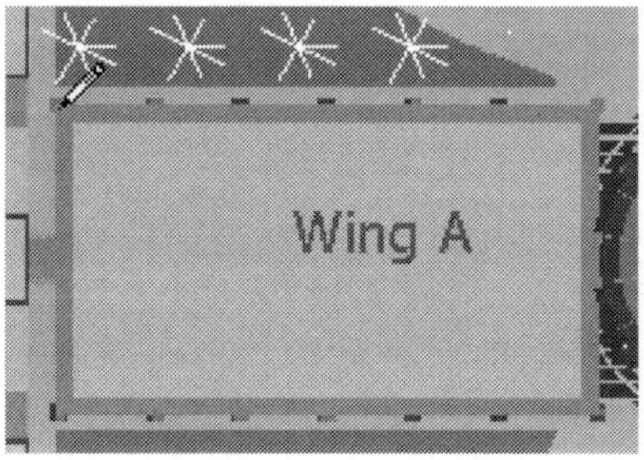

***Figure 3-17** Placing the **Tag** cursor*

***Figure 3-18** Tag 1 created at the corner*

5. Repeat the procedure followed in steps 1 through 4 and add tags on both the wings, as shown in Figure 3-19.

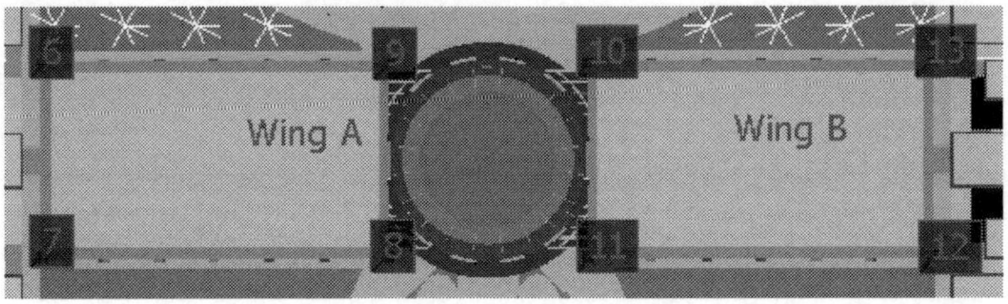

***Figure 3-19** Tags created on both the wings*

Note
*Enter the comment as **Corner point 2**, **Corner point 3**, and so on in the **Add Comment** dialog box.*

Calculating Area

In this section, you will calculate area of both the roofs.

1. Choose the **Area** tool from **Review > Measure > Measure** drop-down.

2. Place the cursor on the corner **6** in **Wing A** and click when the snap to vertex cursor appears.

3. Similarly, click on the corners **7**, **8**, and **9**; the calculated area is displayed in the Scene View, as shown in Figure 3-20.

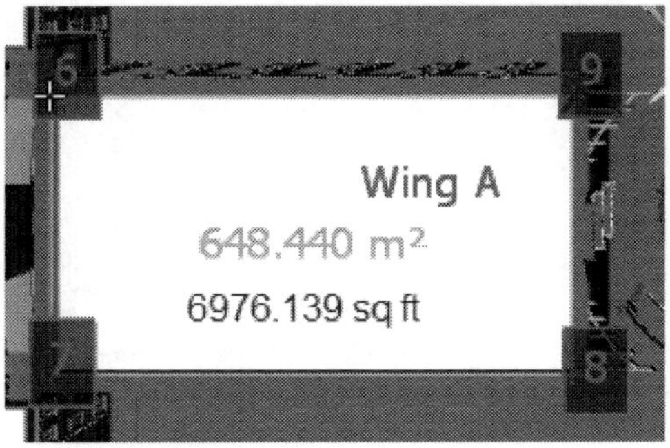

Figure 3-20** Area of roof **WingA

Note
Right-click to change the location of cursor while measuring the area.

4. Similarly, click on the four corners **10**, **11**, **12**, and **13**; the calculated area is displayed, which is almost equal to the area of **Wing A**.

Saving the Project

In this section, you will save the project.

1. To save the project with the current view, choose **Save As** from the Application Menu; the **Save As** dialog box is displayed.

2. Browse to *nws_2024\c03_nws_2024_tut* folder and enter **c03_navisworks_2024_tut02** in the **File name** edit box.

3. Next, select **Navisworks File Set (*.nwf)** file format from the **Save as type** drop-down list, and choose the **Save** button; the project is saved.

Tutorial 3 Finding and Saving Items as Search Sets

In this tutorial, you will open the file *c03_navisworks_2024_tut3.nwf* and then find items based on their types and properties. After finding items, you will save them as search sets.

(Expected time: 20 min)

The following steps are required to complete this tutorial:

a. Open the file *c03_navisworks_2024_tut3.nwf*.
b. Display the **Find Items** and **Sets** windows.
c. Specify the search criteria to find the items and save them as search set.
d. Save the project.

Opening the File

In this section, you will open the file in Navisworks.

1. Choose the **Open** button from the Quick Access Toolbar; the **Open** dialog box is displayed.

2. In this dialog box, browse to the following location: *C:\nws_2024\c03_nws_2024_tut*.

3. Select **Navisworks File Set (*.nwf)** from the **Files of type** drop-down list.

4. Select the file *c03_navisworks_2024_tut3* from the displayed list of files; the file name is displayed in the **File name** edit box.

5. Choose the **Open** button from the dialog box; the model is displayed in the Scene View.

Displaying the Dockable Windows

In this section, you will invoke the **Find Items** and **Sets** windows and dock them on the left side of the interface.

1. Choose the **Find Items** tool from **Home > Select & Search** panel; the **Find Items** window is displayed.

2. Place the cursor on the title bar of the window and double-click; the window is docked at the left side of the interface.

3. Choose the **Manage Sets** option from **Home > Select & Search > Sets** drop-down; the **Sets** window is displayed.

4. Place the cursor on the Title Bar in the **Sets** window, press and hold the left mouse button, drag the **Sets** window and over the Title Bar of the **Properties** window; the two windows are grouped together, refer to Figure 3-21.

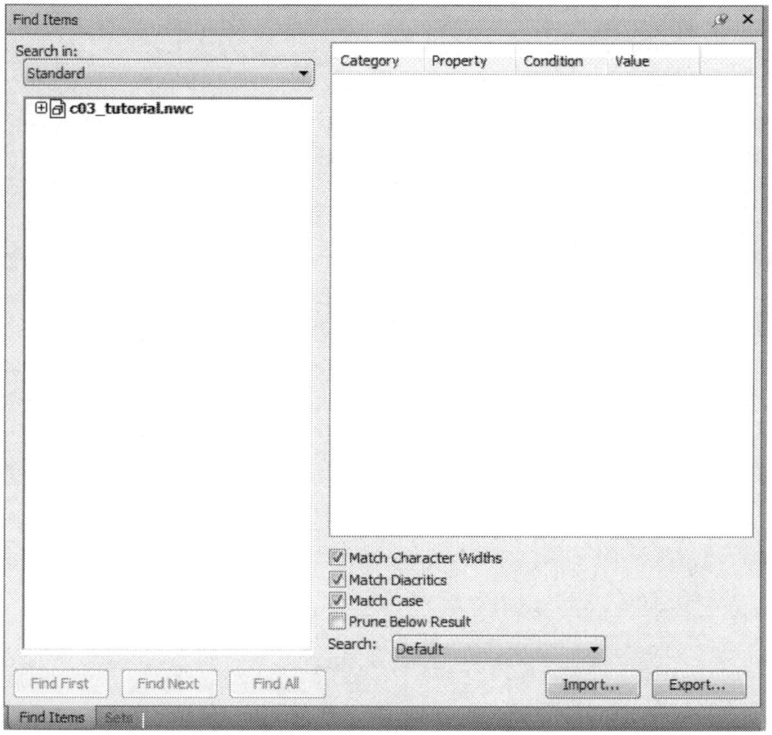

Figure 3-21 *Two windows grouped together*

Finding Items and Saving Items

In this section, you will specify the search criteria to find the items and then save them as search sets.

1. In the **Find Items** window, click in the **Category** column; a drop-down list is displayed. Select **Item** from the drop-down list. Similarly, select **Type**, **Contains**, and **Columns** from the **Property**, **Condition**, and **Value** columns, respectively, refer to Figure 3-22.

2. Choose the **Find All** button; the columns are selected and highlighted in the Scene View.

3. Choose the **Hide Unselected** tool from **Home > Visibility** panel; all the selected columns are displayed, refer to Figure 3-23.

4. Choose the **Unhide All** tool from **Home > Visibility** panel; the whole model will be displayed.

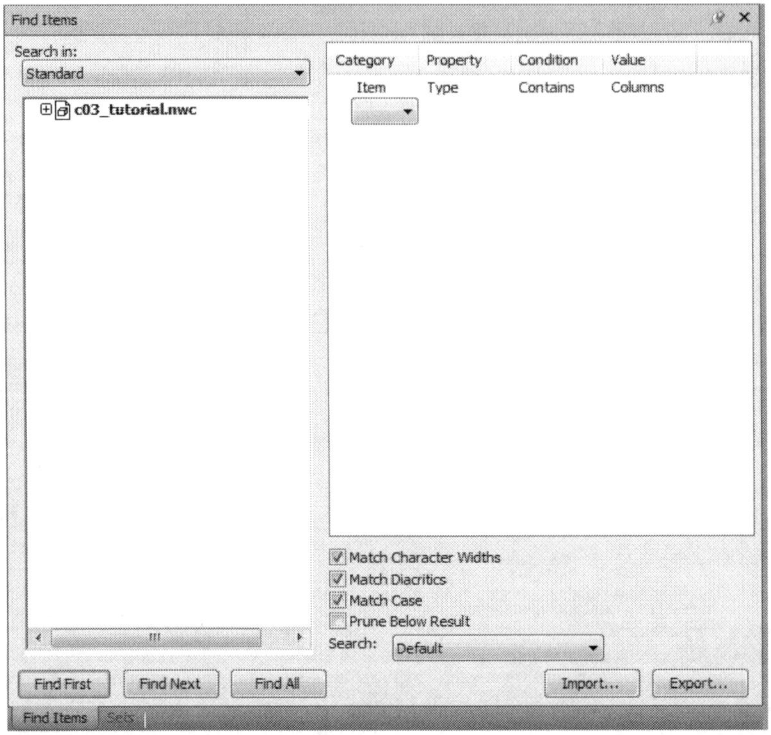

Figure 3-22 *The* ***Find Items*** *window displaying various settings for the search operation*

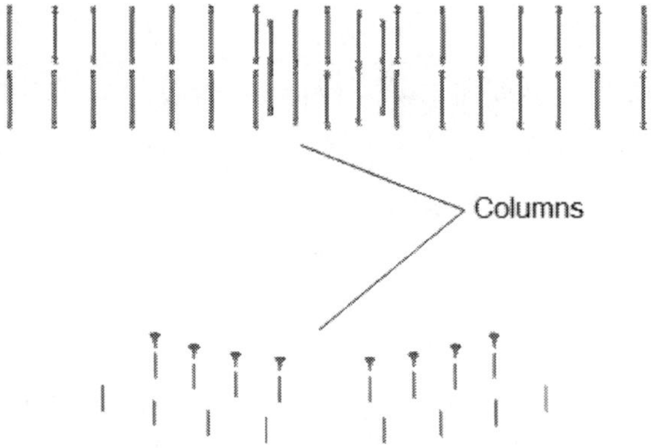

Figure 3-23 *Columns displayed in Scene View*

5. Choose the **Sets** tab in the grouped window; the **Sets** window is displayed. Choose the **Save Search** button; a search set is created. Right-click on the created search set; a shortcut

menu is displayed. Choose the **Rename** option from the menu and name the search set as **Columns** and press ENTER.

6. Choose the **Find Items** tab to display the **Find Items** window. Now, specify a different search criteria, as shown in Figure 3-24.

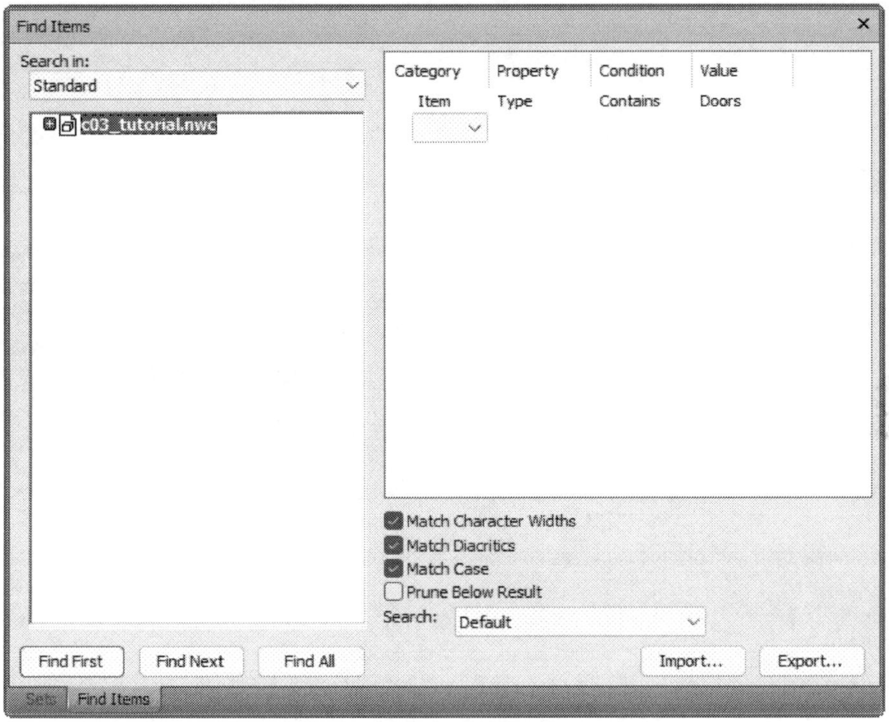

Figure 3-24 *The* ***Find Items*** *window displaying the search criteria*

7. Choose the **Find All** button; objects that have the Doors material assigned are highlighted in the Scene View.

8. Repeat the procedure followed in step 5 and save the selected objects as search set with the name **Doors** and press ENTER.

Saving the Project

In this section, you will save the project.

1. To save the project with the current view, choose **Save As** from the Application Menu; the **Save As** dialog box is displayed.

2. Browse to *nws_2024/c03_nws_2024_tut* folder and enter ***c03_navisworks_2024_tut03*** in the **File name** edit box.

3. Next, select **Navisworks File Set (*.nwf)** file format from the **Save as type** drop-down list and then choose the **Save** button; the project is saved.

SELF-EVALUATION TEST

Answer the following questions and then compare them to those given at the end of this chapter:

1. The **Selection Tree** is invoked from the __________ panel of the **Home** tab.

2. **Selection Inspector** contains a list of all the __________ objects.

3. The __________ option is used to reverse the selection.

4. The properties of an item in the model can be analyzed using the __________ window.

5. The __________ gizmo is used to move the selected item.

6. The **Select** tool is used to select a particular item in a model. (T/F)

7. The **Quick Find** tool is used to analyze the properties of an item in the model. (T/F)

8. In the **Properties** window, you cannot add a custom property to a selected item. (T/F)

9. You can invoke the transformation gizmo from the **Animation** tab. (T/F)

10. The **Accumulate** option is used to calculate the area. (T/F)

REVIEW QUESTIONS

Answer the following questions:

1. Which of the following tabs is part of the **Selection Tree** window?

   a) **Standard** b) **Compact**
   c) **Properties** d) All of these

2. Which of the following tools is not a selection tool?

   a) **Same Name** b) **Same Type**
   c) **Same Timeliner** d) **Require**

3. Which of the following tools is used to control the visibility of objects in the model?

   a) **Append** b) **Merge**
   c) **Find Items** d) None of these

4. Which of the following tools is not a transformation tool?

   a) **Rotate** b) **Scale**
   c) **Move** d) **Text**

5. Which of the following tools is not used as the **Measure** tool?

   a) **Point to Point** b) **Area**
   c) **Add Tag** d) **Angle**

6. The **Measure** tools are invoked from the **Item Tools** tab. (T/F)

7. The **Point to Multiple Points** tool is used to calculate the total distance between multiple points. (T/F)

8. The **Add Link** tool is used to customize the appearance of a model. (T/F)

9. The **Sets** window contains only a list of saved viewpoints. (T/F)

10. The **Convert to Markups** tool is used to convert measurements to markups. (T/F)

EXERCISE

Exercise 1 Measuring Height

Download the *c03_navisworks_2024_ex1* file from *https://www.cadsofttech.com*. The path of the file is as follows: *Books > Civil/GIS Books > Autodesk Navisworks > Exploring Autodesk Navisworks 2024 for Novices (Learn By Doing)* and then calculate the height of the stairs above the ground. Create tags across the four corners of the wall and calculate area of the wall. Figure 3-25 shows the Residence Building model to be used in this exercise. **(Expected time : 30 min)**

The following steps are required to complete the exercise:

1. Open the file *c03_navisworks_2024_ex1*.
2. Display the saved views.
3. Calculate distance between point 1 and 2 using the **Perpendicular Lock** tool.
4. Create tags at the corners of the wall.
5. Calculate area of the wall.
6. Save the project.
7. Name of the file to be saved is *c03_navisworks_2024_ex01*.

Figure 3-25 The Residence Building

Answers to Self-Evaluation Test

1. **Select & Search**, 2. selected, 3. **Invert Selection**, 4. **Properties**, 5. **Move**, 6. T, 7. F, 8. F, 9. F, 10. F.

Chapter 4

Viewpoints, Sections, and Animations

Learning Objectives

After completing this chapter, you will be able to:

- *Work with viewpoints*
- *Create cross-sections for reviewing models*
- *Create and share animations*

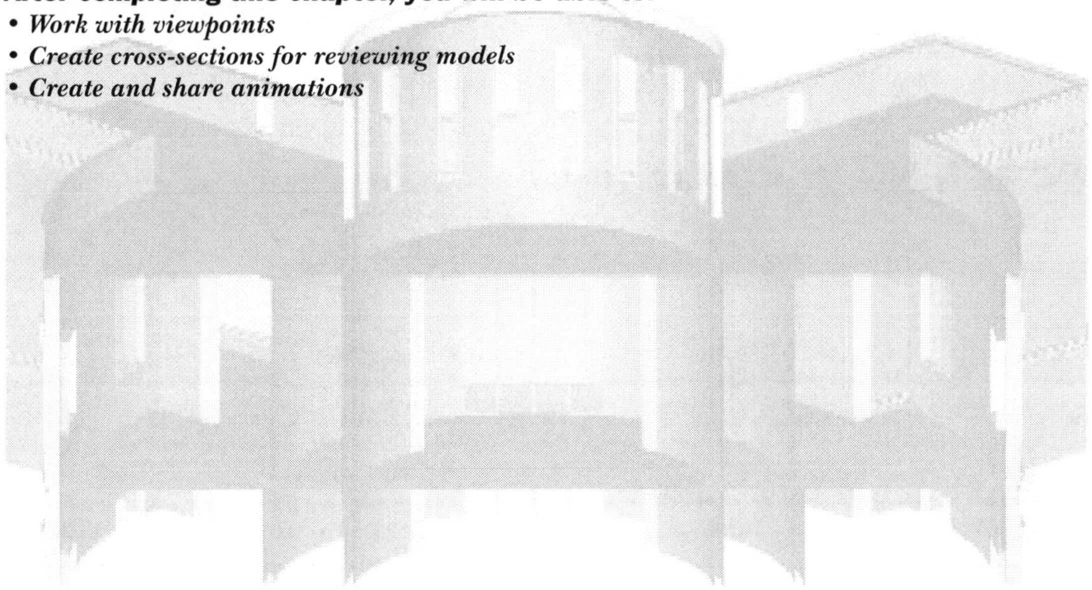

INTRODUCTION

In the previous chapter, you learned to select objects, control object attributes, and use review tools. In this chapter, you will learn various methods of saving and using viewpoints, creating cross sections, and the basic concepts of animation. All these will be discussed sequentially in this chapter.

WORKING WITH VIEWPOINTS

In Navisworks, you can save views as a viewpoint in a project. The viewpoints are snapshots of the views created in a scene. The saved viewpoints are used to manage and review the changes in a project.

Saving Viewpoints

In a project, sometimes you need to recall previous views. You can save model views by saving viewpoints. Before saving you need to set orientations and add mark up, tags, and comments.

Organizing Viewpoints

In the **Saved Viewpoints** window, you can arrange the saved views in a systematic manner by creating folders and sub folders. This helps you to easily locate them later.

Editing Viewpoints

You can configure several attributes of the saved viewpoints such as camera position, field of view, linear speed, and angular speed.

Sharing Viewpoints

While working in a project, you can share the created viewpoints with other users. To use the viewpoints from another model, you need to import it via XML file.

SECTIONING IN Navisworks

In Navisworks, sectioning is used to create cross-section views of the model so that you can get a clear picture of the internal view of the model. Using this feature, you can create different views in a model and can link them together.

Creating Section Views Using Planes Mode

In the Planes mode, you can enable a maximum of six sectional planes for creating the cross-section views of the model.

Transforming Section Planes

The section planes can be transformed using transformation gizmos such as **Move** and **Rotate**. They can also be transformed numerically. Next, you will learn to transform these section planes using different methods.

Linking Section Planes

In a project, you can link all the section planes together using the **Link Section Planes** option to move them as one section.

Creating Section Views Using the Box Mode

In the Box mode, a section box is created around the model. Using this section box, you can focus on a specific area of the model by limiting its geometry. You can clip the unwanted portion of the model in the Scene View by using the Section Box mode.

Transforming Section Box

The section box can be transformed using transformation gizmos such as **Move**, **Rotate**, and **Scale**, and it can also be transformed numerically.

ANIMATION IN NAVISWORKS

In Navisworks, you can create animations by using two methods:

1) Viewpoint animation
2) Object animation

The Viewpoint animation is an efficient way of creating animation by recording views while navigating through a model or by saving specific views of the model.

In the Object animation, after animating objects, you can interact with them by adding scripts such as opening doors or closing doors. In this animation, when you move an object to the desired location, many keyframes will be generated between the start point and end point, thus making the transitions smooth and speeding up the process of animation.

Creating Viewpoint Animations By Recording

In Navisworks, you can create viewpoint animation while navigating through a model by using the **Record** tool. It will record each movement while navigating, and create frames for each movement. All these frames are saved in the animation folder in the **Saved Viewpoints** window. When you play the animation, all these frames will be played in a sequence.
You can also create Viewpoint Animations Frame By Frame.

Playing Viewpoint Animations

After creating an animation, you can play it by selecting it from **Animation > Playback > Animations** drop-down, refer to Figure 4-1.

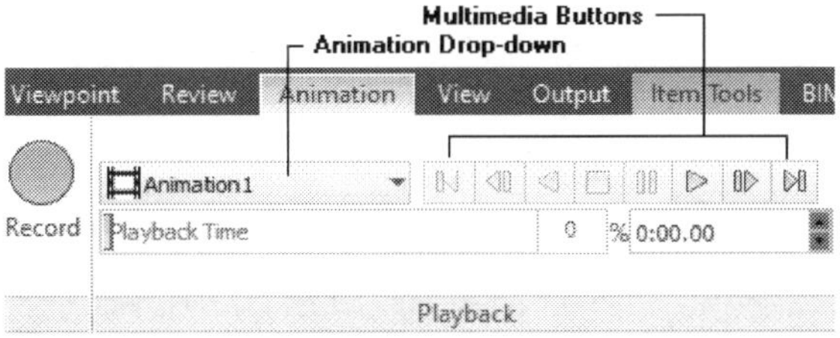

Figure 4-1 *The* ***Playback*** *panel in the* ***Animation*** *tab*

Combining Multiple Animations

In Navisworks, multiple animations can be combined and played together as a single animation. These animations are created either by using the **Record** tool or they can be created frame by frame.

Editing Viewpoint Animations

After creating an animation, you can edit the playback duration. You can also edit the smoothness of the animation playback by reducing the jerk.

Sharing Animations

In Navisworks, you can share the animation by using the export feature. To use it, choose the **Export Animation** tool from the **Export** panel in the **Animation** tab; the **Animation Export** dialog box will be displayed, as shown in Figure 4-2.

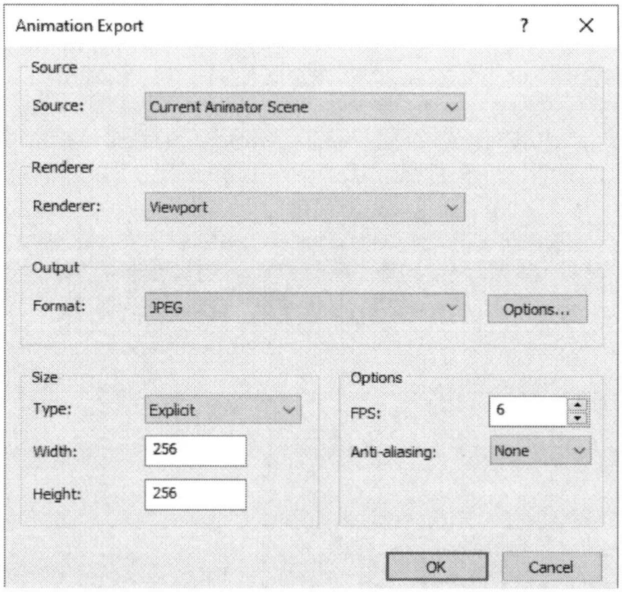

Figure 4-2 *The* ***Animation Export*** *dialog box*

TUTORIALS

General instructions for downloading tutorial files:

1. Download the *c04_navisworks_2024_tut* zip file for the tutorial from *https://www.cadsofttech.com*. The path of the file is as follows: ***Books > Civil/GIS Books > Autodesk Navisworks > Exploring Autodesk Navisworks 2024 for Novices (Learn By Doing) > Tutorial Files.***

2. Now, save and extract the downloaded folder at the following location:
C:\ nws_2024\c04_nws_2024_tut

Tutorial 1 Creating Viewpoints

In this tutorial, you will open the *c04_navisworks_2024_tut1* file, create viewpoints, and save them in the **Saved Viewpoints** window. **(Expected time: 20 min)**

The following steps are required to complete this tutorial:

a. Open the *c04_navisworks_2024_tut1.nwf* file.
b. Display the **Saved Viewpoints** window.
c. Display the 3D view of the model.
d. Create viewpoints.
e. Save the project.

Opening the Model

In this section, you will open the model created in Revit software.

1. Choose the **Open** button from the Quick Access Toolbar; the **Open** dialog box is displayed.

2. In this dialog box, browse to the following location:
   C:\nws_2024\c04_nws_2024_tut

3. Select the **Navisworks File Set (*.nwf)** file format from the **Files of type** drop-down list.

4. Next, select *c04_navisworks_2024_tut1* from the displayed list of files in the dialog box, and choose the **Open** button; the model is displayed in the Scene View, as shown in Figure 4-3.

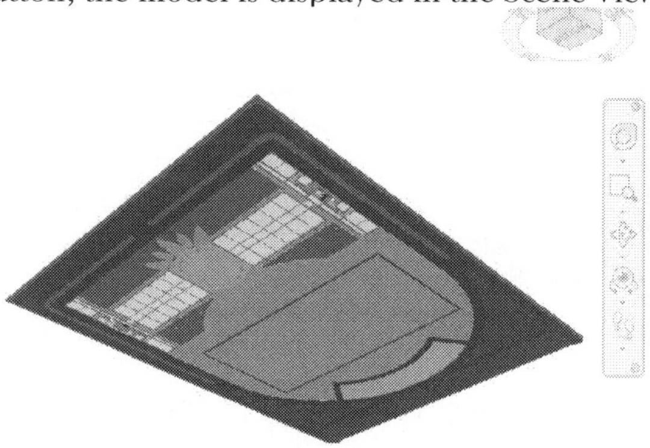

Figure 4-3 *Model opened in Navisworks*

Displaying the Saved Viewpoints Window

In this section, you will display the **Saved Viewpoints** window.

1. Choose the **Saved Viewpoints Dialog Launcher** button from the **Save, Load & Playback** panel in the **Viewpoint** tab; the **Saved Viewpoints** window is displayed.

Displaying the 3D View of the Model

In this section, you will display the 3D view of the model.

1. Expand the **3D View** node in the **Saved Viewpoints** window.

2. Select the **North Elevation** viewpoint in the **3D View** node; the current view changes and the north elevation view of the model is displayed, as shown in Figure 4-4.

Figure 4-4 *North elevation of the model*

Note

*If an avatar is displayed with the **North Elevation** view then clear the **Third Person** check box from the **Realism** drop-down in the **Viewpoint** tab.*

Creating Viewpoints

In this section, you will create viewpoints.

1. Invoke the **Orbit** tool from the Navigation Bar; the **Orbit** cursor appears on the screen.

2. Press and drag the left mouse button and rotate the current view to align it, as shown in Figure 4-5.

Figure 4-5 *The model after rotating*

3. Invoke the **Zoom** tool from the Navigation Bar; the Zoom cursor appears on the screen.

4. Place the **Zoom** cursor on the model, as shown in Figure 4-6.

***Figure 4-6** Placing the **Zoom** cursor on the model*

5. Enlarge the viewing scale of the model by clicking and dragging the mouse forward, as shown in Figure 4-7.

***Figure 4-7** The model after zooming*

6. Again, place the **Zoom** cursor on the door, as shown in Figure 4-8.

***Figure 4-8** Cursor placed on the door*

7. Enlarge the viewing scale of the door by dragging the mouse forward, as shown in Figure 4-9.

Figure 4-9 *Partial view of the door after zooming*

8. Right-click in the **Saved Viewpoints** window; a shortcut menu is displayed.

9. Choose the **New Folder** option from the menu; a folder is created and an edit box is displayed in the **Saved Viewpoints** window.

10. Rename the folder created as **Saved Views** and press ENTER.

11. Next, right-click in the **Saved Views** folder; a shortcut menu is displayed.

12. Choose the **Save Viewpoint** option from the displayed menu; a viewpoint is created. Rename the saved viewpoint as **Measurements** and press ENTER.Figure 4-10 shows the **Measurement** viewpoint saved in the **Saved Viewpoints** window.

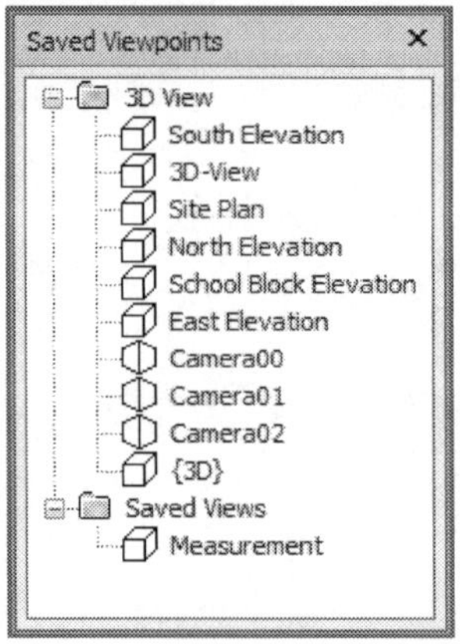

Figure 4-10 *The* ***Measurement*** *viewpoint saved in the* ***Saved Viewpoints*** *window*

13. Next, using the **Zoom** tool, zoom out the model by clicking and dragging the mouse backward, as shown in Figure 4-11.

Figure 4-11 *Model after zooming out*

14. Next, move and place the cursor on the **Top** face of the ViewCube; the top face of the ViewCube is highlighted.

15. Click on the **Top** face of the ViewCube; the view of the model changes to the top view, as shown in Figure 4-12.

Figure 4-12 *Top view of the model*

Note

You can zoom out the model to get a view similar to the one shown in Figure 4-13.

16. Now, right-click in the **Saved Views** folder in the **Saved Viewpoints** window; a shortcut menu is displayed.

17. Choose the **Save Viewpoint** option from the menu.

18. Repeat the procedure followed in steps 11 and 12 and save the viewpoint with the name **Area**. Figure 4-31 shows the **Saved Viewpoints** window with the **Area** viewpoint.

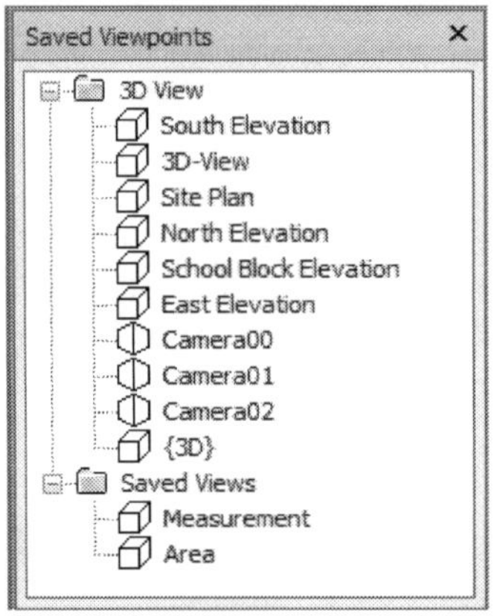

***Figure 4-13** The **Saved Viewpoints** window*

19. Select the **3D View** in the **Saved Viewpoints** window; the view is displayed in the Scene View. Now, close the **Saved Viewpoints** window.

Saving the Project

In this section, you will save the project.

1. Choose **Save As** from the Application Menu; the **Save As** dialog box is displayed.

2. Browse to *nws_2024\ c04_nws_2024_tut* folder and enter **c04_navisworks_2024_tut01** in the **File name** edit box. Choose the **Navisworks File Set (*.nwf)** file format from the **Save as type** drop-down list, and choose the **Save** button; the project is saved.

Tutorial 2 Using Planes Mode

In this tutorial, you will open the *c04_navisworks_2024_tut2* file and create cross-section of the model in the Planes mode to get the inside view. **(Expected time: 30 min)**

The following steps are required to complete this tutorial:

a. Open the *c04_navisworks_2024_tut2.nwf* file.
b. Enable sectioning.
c. Create sections using planes mode and various transforming tools.
d. Save the project.

Opening the Model

In this section, you will open the model created in Revit software.

1. Choose the **Open** button from the Quick Access Toolbar; the **Open** file dialog box will be displayed on the screen.

2. In this dialog box, browse to the following location: *C:\nws_2024\c04_nws_2024_tut*.

3. Ensure that the **Navisworks File Set (*.nwf)** file format is selected in the **Files of type** drop-down list. Next, select the *c04_navisworks_2024_tut2* from the displayed list of files and choose the **Open** button; the model is displayed in the Scene View, as shown in Figure 4-14.

Figure 4-14 *Model opened in Navisworks*

Enabling Cross-section Mode

In this section, you will enable the cross-section mode of the current view of the model.

1. Choose the **Enable Sectioning** option from the **Sectioning** panel in the **Viewpoint** tab; the **Sectioning Tools** contextual tab is displayed in the ribbon.

Creating Sections Using Planes Mode and Transforming Tools

In this section, you will use the **Planes** tool to create cross-sections of the model.

1. Ensure that the **Planes** tool is chosen from the **Planes** drop-down in the **Mode** panel of the **Sectioning Tools** tab.

2. Ensure that the **Plane 1** option is selected in the **Current: Plane** drop-down list in the **Planes Settings** panel of the **Sectioning Tools** tab.

3. Choose the **Move** tool from the **Transform** panel in the **Sectioning Tools** tab; the section plane is displayed, as shown in Figure 4-15.

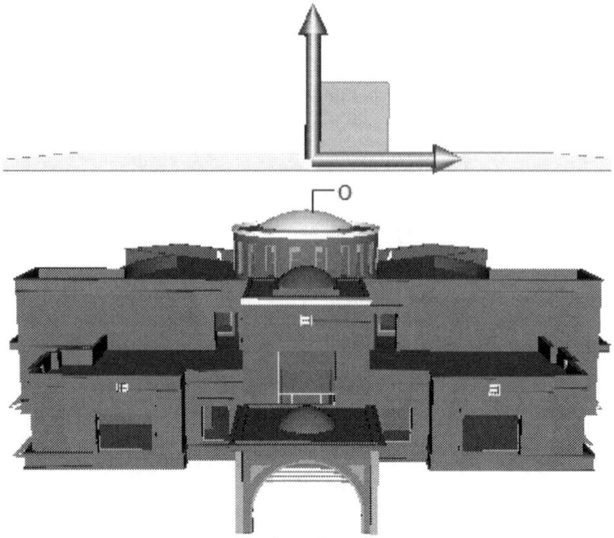

Figure 4-15 Model with the section plane

4. Invoke the **Select** tool from the Quick Access Toolbar.

5. Select the building element marked O, refer to Figure 4-15; the selected object is highlighted in blue.

6. Choose the **Fit Selection** tool from the **Transform** panel in the **Sectioning Tools** tab; the section plane moves to the highlighted building element, as shown in Figure 4-16.

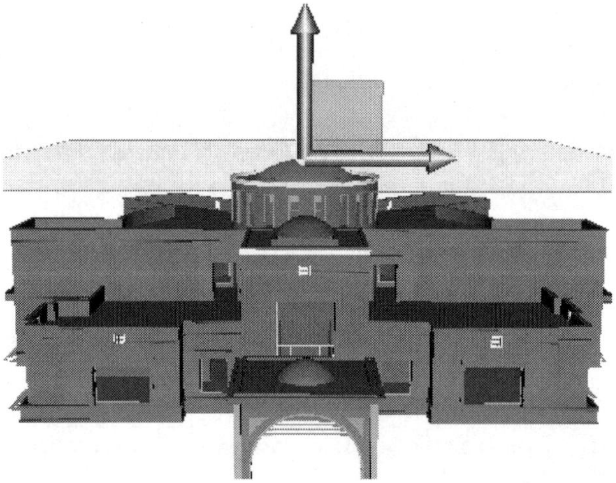

Figure 4-16 Section plane moved to the selected building element

7. Choose the **Left** alignment option from **Sectioning Tools > Plane Settings > Alignment** drop-down list; the section plane changes, as shown in Figure 4-17.

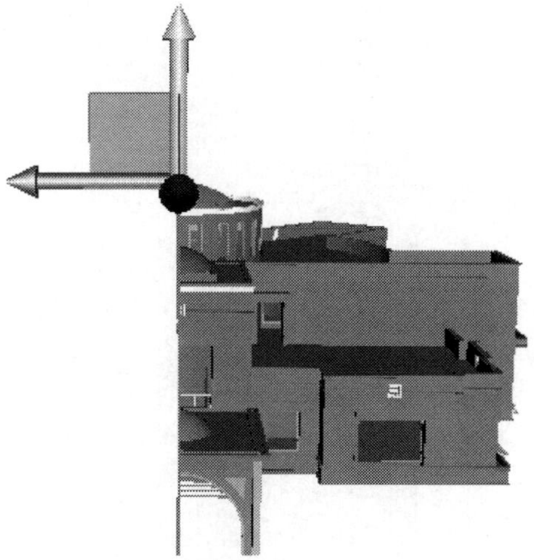

Figure 4-17 *Model with the left aligned section plane*

8. Next, choose the **Rotate** tool from the **Transform** panel in the **Sectioning Tools** tab; the **Rotate** gizmo is displayed, as shown in Figure 4-18.

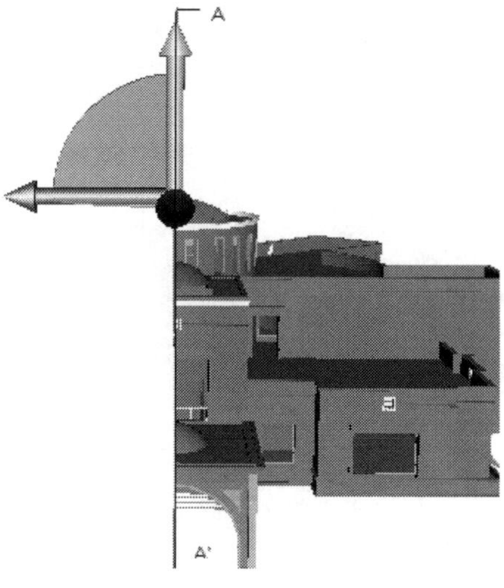

Figure 4-18 *The* ***Rotate*** *gizmo*

9. Place the cursor on the red curve between the Y and Z axis; the curve gets highlighted, as shown in Figure 4-19.

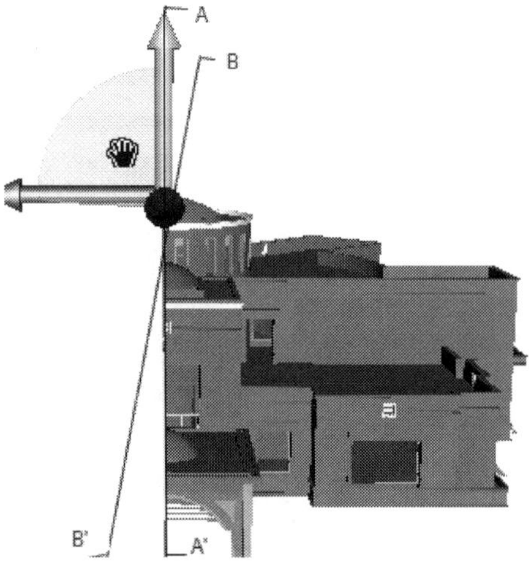

Figure 4-19 *Placing the cursor on the curve*

10. Press and hold the left mouse button, drag the **Rotate** gizmo in the right direction, and release it when the section plane meets the BB' axis, refer to Figure 4-19; the section plane rotates, as shown in Figure 4-20.

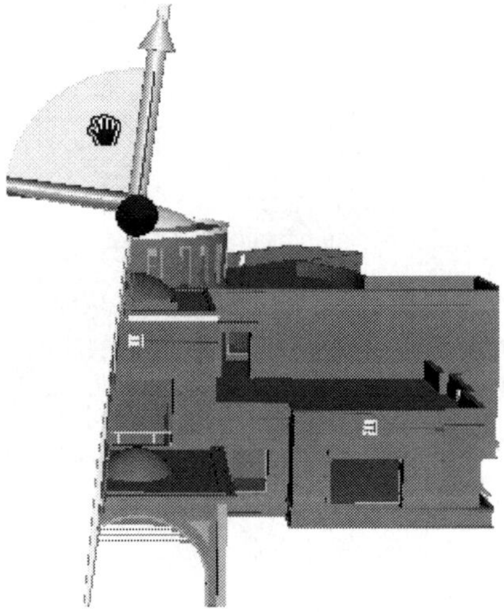

Figure 4-20 *The model after rotating*

11. Deselect the **Rotate** tool from the **Transform** panel in the **Sectioning Tools** tab; the **Rotate** gizmo and section plane disappears, as shown in Figure 4-21.

12. Next, invoke the **Orbit** tool from **Viewpoint > Navigate > Orbit** drop-down; the Orbit cursor appears in the Scene View.

13. Press and hold the left mouse button, drag the cursor in the right direction and release the cursor when the section view of the model looks similar to the one shown in Figure 4-22.

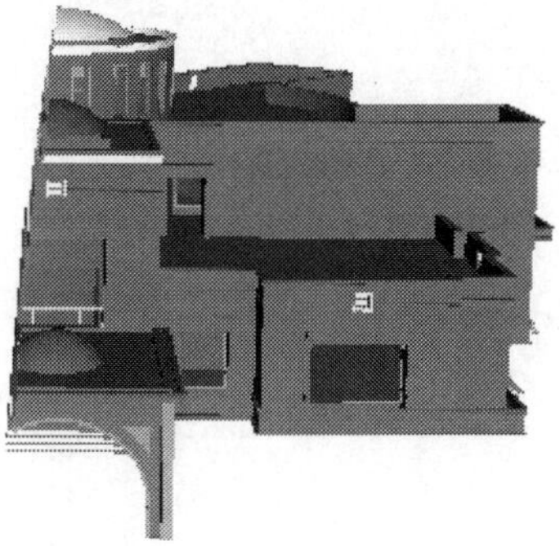

Figure 4-21 *Rotated and sectioned view of the model*

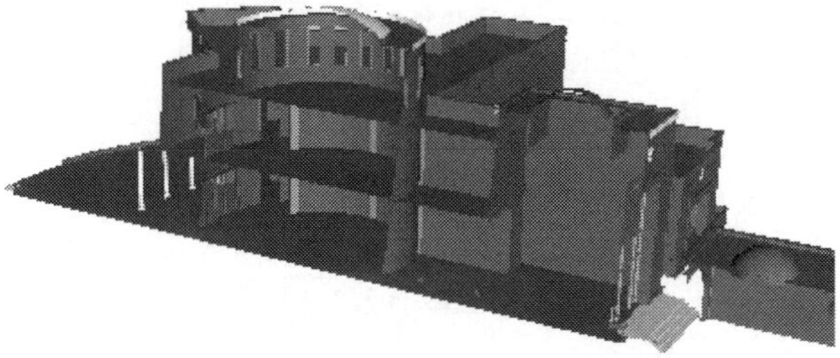

Figure 4-22 *Section view of the model*

14. Now, choose the **Save Viewpoint** option from the **Save** panel of the **Sectioning Tools** tab; the **Saved Viewpoints** window is displayed in the Scene View.

15. Enter **Section View** in the edit box displayed in the **Saved Viewpoints** window, and press ENTER; the view is saved as Section View.

16. Close the **Saved Viewpoints** window and choose the **Enable Sectioning** button in the **Sectioning** panel of the **Viewpoint** tab; the sectioned model is displayed, as shown in Figure 4-23.

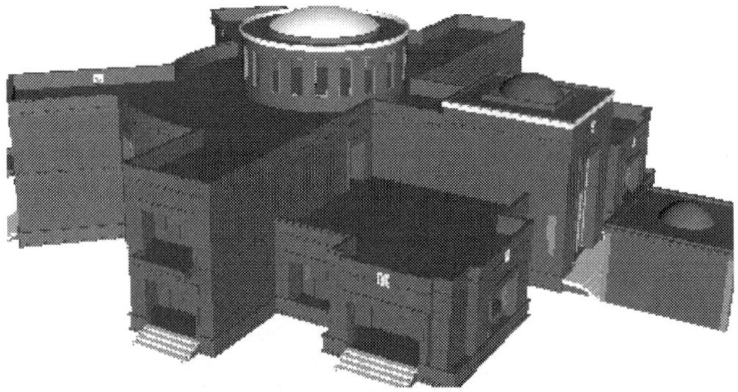

Figure 4-23 *The sectioned model displayed in the Scene View*

Saving the Project

In this section, you will save the project.

1. Choose **Save As** from the Application Menu; the **Save As** dialog box is displayed.

2. Browse to *nws_2024\ c04_nws_2024_tut* folder and enter **c04_navisworks_2024_tut02** in the **File name** edit box. Choose **Navisworks File Set (*.nwf)** file format from the **Save as type** drop-down list, and then choose the **Save** button; the project is saved.

Tutorial 3 Using Section Box Mode

In this tutorial, you will open the *c04_navisworks_2024_tut3* file and use the Section Box mode to focus on specific area of the model. **(Expected time: 30 min)**

The following steps are required to complete this tutorial:

a. Open the file *c04_navisworks_2024_tut3* in Navisworks.
b. Enable Sectioning.
c. Use section box and transforming tools.
d. Save the viewpoint.
e. Save the project.

Opening the Model

In this section, you will open the model created in Revit software.

1. Choose the **Open** button from the Quick Access Toolbar; the **Open** dialog box is displayed.

2. In this dialog box, browse to the following location:
C:\nws_2024\c04_nws_2024_tutorial.

3. Ensure that the **Navisworks File Set (*.nwf)** option is selected in the **Files of type** drop-down list.

4. Select the file **c04_navisworks_2024_tut3** from the displayed list of files; the file name appears in the **File name** edit box.

5. Choose the **Open** button in the dialog box; the model is displayed in the Scene View, as shown in Figure 4-24.

Figure 4-24 *Model opened in Navisworks*

Enabling Cross-Section Mode

In this section, you will enable the cross-sectioning mode of the current view of the model.

1. Choose the **Enable Sectioning** option from the **Sectioning** panel in the **Viewpoint** tab; the **Sectioning Tools** contextual tab is displayed in the ribbon.

Using the Section Box Mode and Transforming Tools

In this section, you will use the **Section Box** mode option for the cross-sectioning of the model.

1. Choose the **Box** tool from the **Sectioning Tools > Mode > Plane** drop-down; the Section Box is displayed along with the **Move** gizmo in the Scene View, as shown in Figure 4-25.

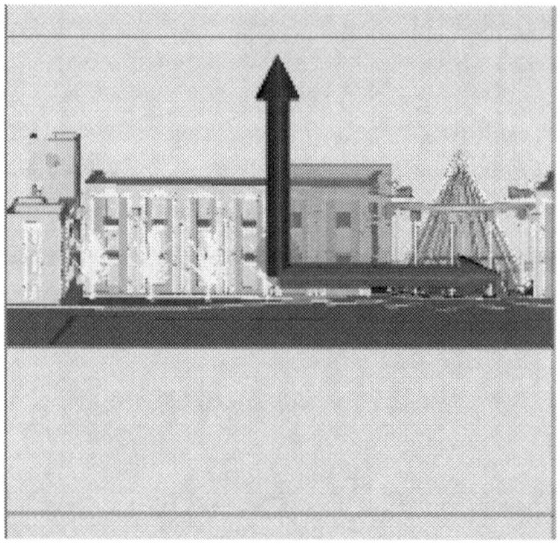

Figure 4-25 *The Section Box with the **Move** gizmo*

Note

*If the Section Box and the **Move** gizmo are not displayed, then choose the **Move** tool from the **Transform** panel in the **Sectioning Tools** tab. Also, note that the direction of the gizmo may vary.*

2. Place the cursor on the red arm (x-axis); the axis is highlighted and a hand symbol appears, as shown in Figure 4-26.

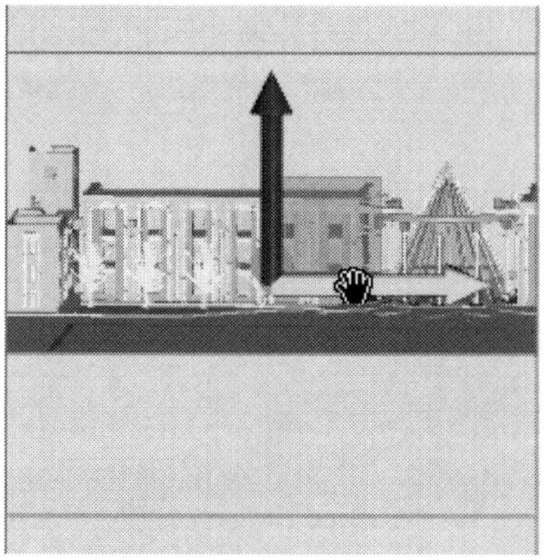

Figure 4-26 *Placing the cursor on the red arm of the gizmo*

3. Press and hold the left mouse button and then drag the gizmo in the left direction. Release the mouse button when the model looks similar to the one shown in Figure 4-27.

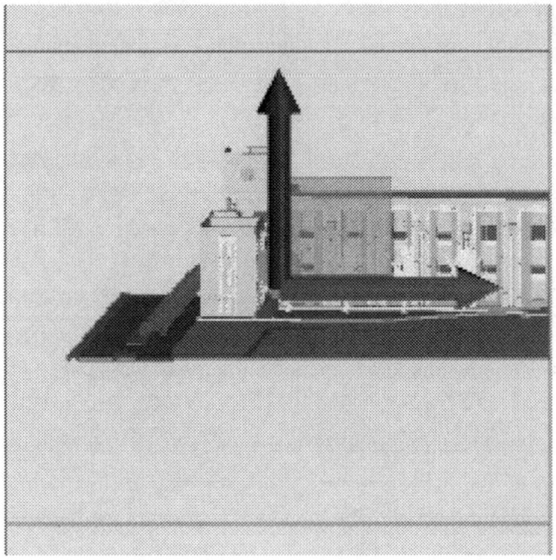

Figure 4-27 *Model after dragging the gizmo in the left direction*

4. Place the cursor on the green arm; the axis is highlighted and a hand symbol appears, as shown in Figure 4-28.

Note

If the green arrow is not visible, invoke the ***Orbit*** *tool and rotate the model in the right direction.*

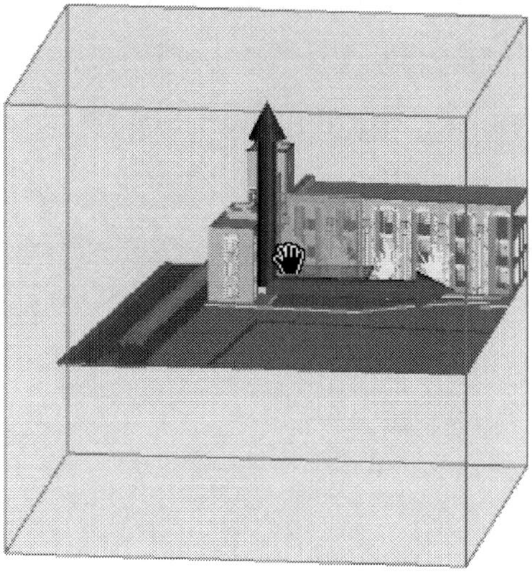

Figure 4-28 *Placing the cursor on the green arm of the* ***Move*** *gizmo*

5. Press and hold the left mouse button and then drag the gizmo in backward direction. Release the mouse button when the model looks similar to the one shown in Figure 4-29.

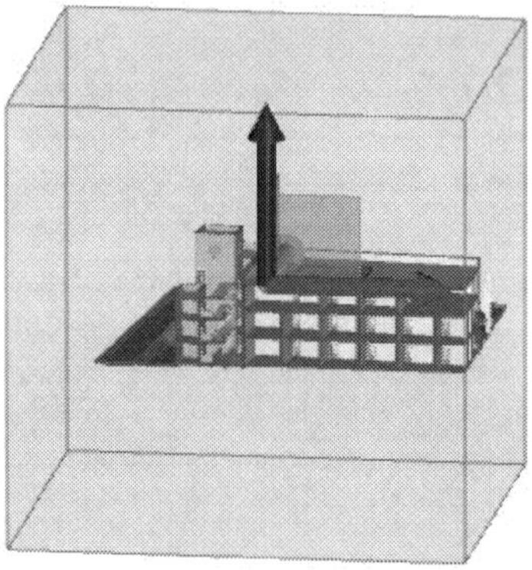

Figure 4-29 *Model after dragging the gizmo in the backward direction*

6. Choose the **Scale** tool from the **Transform** panel in the **Sectioning Tools** tab; the **Scale** gizmo appears inside the **Section Box**, as shown in Figure 4-30.

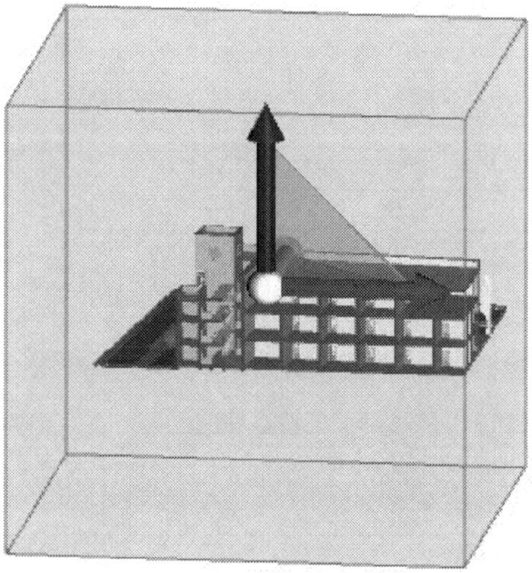

Figure 4-30 *The* **Scale** *gizmo inside the section box*

7. Place the cursor on the green triangle (triangle between X and Z axes); the triangle gets highlighted, as shown in Figure 4-31.

Note
Using the curve between the X and Z axes will enlarge the section box along the Y axis.

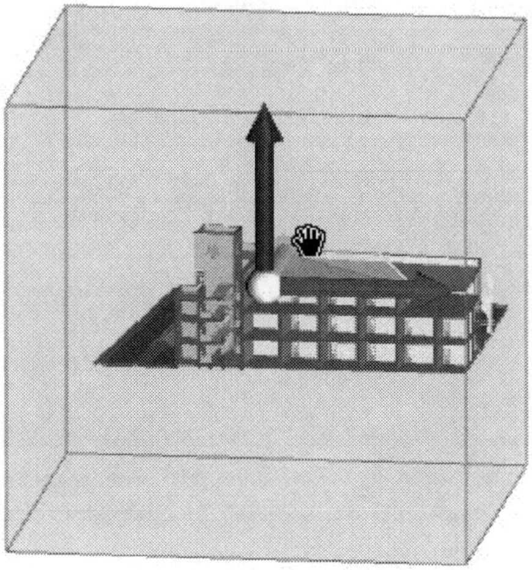

Figure 4-31 *Placing the cursor on the green triangle of the* ***Scale*** *gizmo*

8. Press and hold the left mouse button and then drag the cursor in backward direction. Release the button when the model looks similar to the one shown in Figure 4-32.

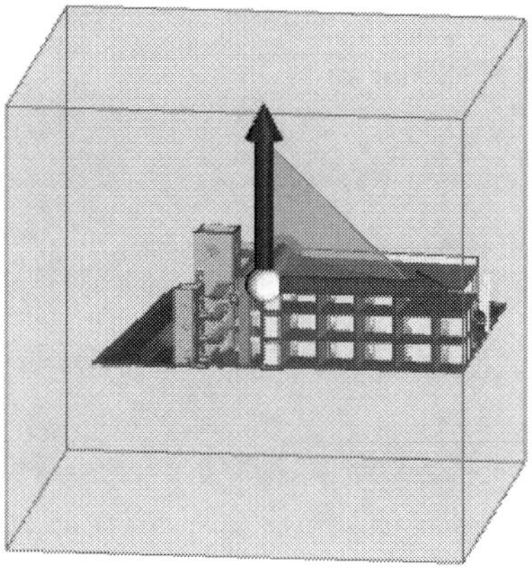

Figure 4-32 *Model after using the* ***Scale*** *gizmo*

9. Choose the **Scale** tool from the **Transform** panel of the **Sectioning Tools** tab; the section box and the **Scale** gizmo disappear and the model is displayed, as shown in Figure 4-33.

Figure 4-33 *The model after the* ***Scale*** *gizmo disappears*

10. Choose the **Zoom Window** tool from the **Zoom** drop-down in the **Navigation Bar**; the shape of the cursor changes.

11. Place the cursor at the corner in the sectioned model, as shown in Figure 4-34.

12. Press and hold the left mouse button and then drag the cursor to draw a window, as shown in Figure 4-34.

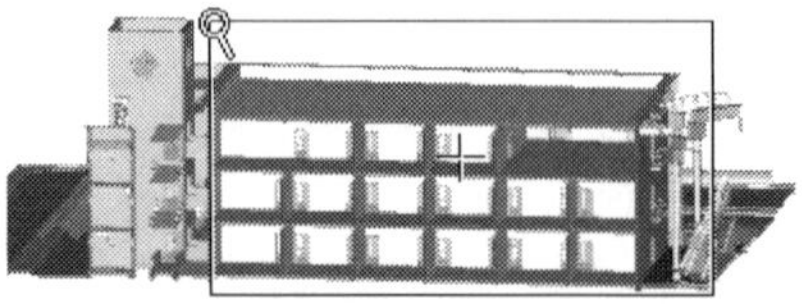

Figure 4-34 *Placing the* ***Zoom Window*** *cursor*

Figure 4-35 *Drawing the zoom window*

13. Release the mouse button; the model is enlarged, as shown in Figure 4-36.

Figure 4-36 *The enlarged view of the sectioned model*

Saving the Viewpoint

In this section, you will save the current sectioned viewpoint.

1. Choose the **Save Viewpoint** button from the **Save** panel of the **Sectioning Tools** tab; the current view is saved in the **Saved Viewpoints** window.

2. Specify **Section View** in the edit box of the **Saved Viewpoints** window and press ENTER.

Saving the Project

In this section, you will save the project.

1. Choose **Save As** from the Application Menu; the **Save As** dialog box is displayed.

2. Browse to the *nws_2024\ c04_nws_2024_tut* folder and enter **c04_navisworks_2024_tut03** in the **File name** edit box. Then, choose **Navisworks File Set (*.nwf)** file format from the **Save as type** drop-down list. Now, choose the **Save** button to save the project.

SELF-EVALUATION TEST

Answer the following questions and then compare them to those given at the end of this chapter:

1. The saved viewpoints are displayed in the__________ window.

2. The __________ dialog box is used to modify various viewpoint properties.

3. The two sectioning modes available in Navisworks are __________ and __________.

4. The **Link Section Plane** option is available in the __________ panel.

5. The __________ mode is used to focus on a specific area of the model.

6. The **Add Comment** dialog box is used to modify viewpoints. (T/F)

7. The viewpoints are exported from Navisworks to the XML file. (T/F)

8. In the Section Box mode, six sectional cuts are created in the model. (T/F)

9. The **Record** tool is used to record the animation while navigating through a model. (T/F)

10. Navisworks does not allow you to combine multiple animations into a single animation. (T/F)

11. You cannot save any hidden information in a viewpoint. (T/F)

REVIEW QUESTIONS

Answer the following questions:

1. Which of the following tabs contains the **Record** tool?

   a) **Viewpoint** tab b) **Animation** tab
   c) **Home** tab d) None of these

2. Which of the following tools is used for displaying the **Section Plane Settings** window?

   a) Record b) Section Plane Dialog Launcher
   c) Box d) None of these

3. Which of the following options is used as an animation progress indicator?

a) **Playback Time** b) **Percentage**
c) **Viewport** d) None of these

4. Which of the following is a type of animation?

a) Viewpoint Animation b) Object Animation
c) Play Animation d) None of these

5. Which of the following options is used to create six sectional cuts in a model?

a) **Box** b) **Record**
c) **Planes** d) None of these

6. The **Loop Playback** check box is used to play the animation continuously. (T/F)

7. The **Fit Selection** option is used to move the section plane to the selected object. (T/F)

8. The **Rotate** gizmo is not used with section box. (T/F)

9. The **Edit Animation** dialog box is used to make changes in animation. (T/F)

10. The **Link Section Plane** option is used to link all the selected section planes together. (T/F)

EXERCISE

Exercise 1 Creating Sections

Download and open the *c04_navisworks_2024_ex1* file from *https://www.cadsofttech.com*. The path of the file is as follows: ***Books > Civil/GIS Books > Autodesk Navisworks > Exploring Autodesk Navisworks 2024 for Novices (Learn By Doing)***. Figure 4-37 shows the model to be used for this exercise. Section the model using the **Planes** and **Box** tools. **(Expected time: 45 min)**

The steps required to complete this exercise are given below:

1. Open the file *c04_navisworks_2024_ex1*.
2. Adjust the model using the **Orbit** and **Zoom** tools.
3. Enable sectioning.
4. Create planes using the **Planes** mode.
5. Transform section planes using transformation tools.
6. Create sections using the **Box** mode.
7. Transform section box using transformation tools.
8. Save the viewpoints.
9. Save the project as *c04_navisworks_2024_ex01*.

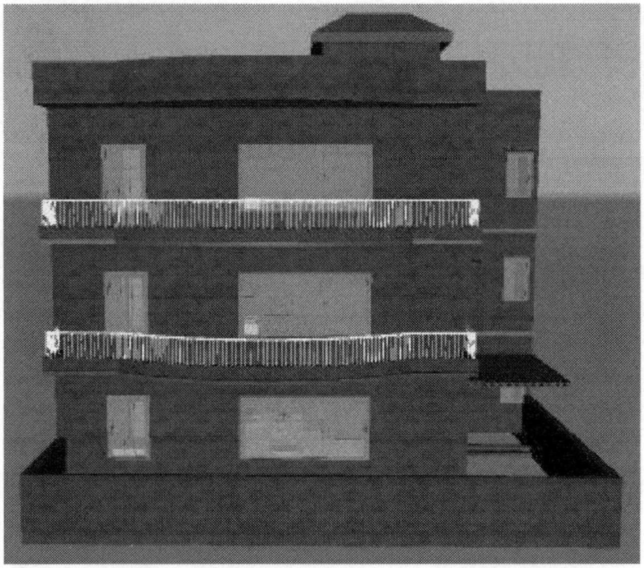

Figure 4-37 *The Residence building*

Exercise 2 Saving Viewpoints

Download and open the *c04_navisworks_2024_ex1* file from *https://www.cadsofttech.com*. The path of the file is as follows:***Books > Civil/GIS Books > Autodesk Navisworks > Exploring Autodesk Navisworks 2024 for Novices (Learn By Doing)***. Save the viewpoints using the **Save Hide/Required Attributes** option. You will observe both the viewpoints. Figures 4-38 and 4-39 show the model and the **Saved Viewpoints** window after saving the viewpoints. **(Expected time: 45 min)**

The steps required to complete this exercise are given below:

1. Open the file *c04_navisworks_2024_ex1*.
2. Invoke the **Saved Viewpoints** window and create two folders with the name Saved Views1 and Saved Views2.
3. Select the **Save Hide/Required Attributes** check box in the **Options Editor** dialog box.
4. Select the wall, hide it, and save it as a viewpoint in the Saved Views1 folder.
5. Name the viewpoint as Wall Hidden.
6. Save a viewpoint with the wall shown and name it as Wall in the Saved Views folder.
7. Clear the **Save Hide/Required Attributes** check box in the **Options Editor** dialog box.
8. Save two viewpoints one with the wall hidden and one with the wall shown in Saved Views2 folder.
9. Compare the differences between the viewpoints.
10. Save the project as *c04_navisworks_2024_ex02*.

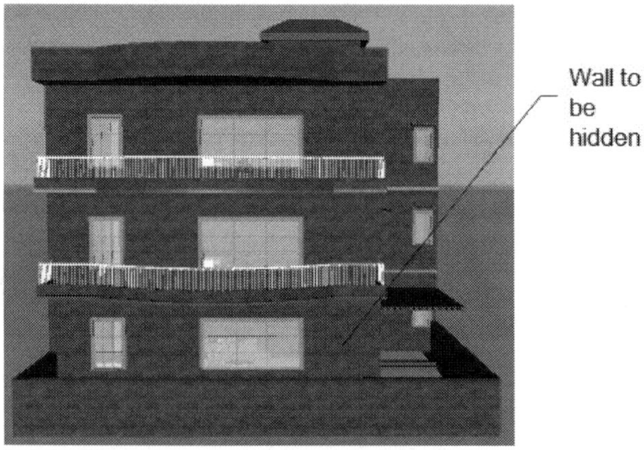

Figure 4-38 Model used in the exercise

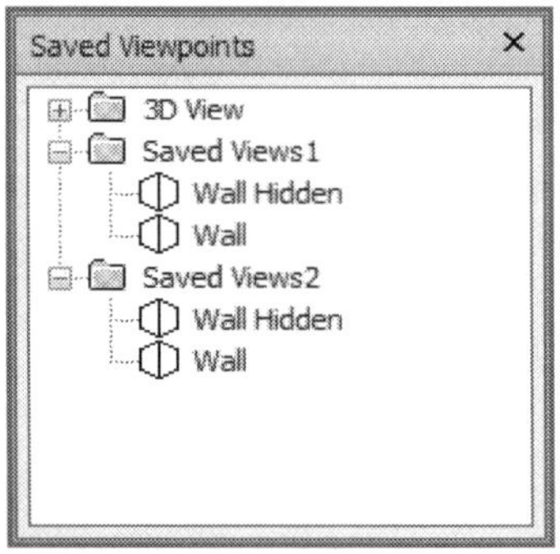

Figure 4-39 Saved viewpoints

Answers to Self-Evaluation Test

1. Saved Viewpoints, **2. Edit Viewpoint**, **3. Planes** and **Box**, **4. Planes Settings**, **5. Box**, **6.** F, **7.** T, **8.** F, **9.** T, **10.** F **11.** F

Chapter 5

TimeLiner

Learning Objectives

After completing this chapter, you will be able to:

- *Understand the concept of TimeLiner*
- *Create tasks in TimeLiner*
- *Attach items to tasks*
- *Arrange tasks in a hierarchy*
- *Configure tasks parameters*
- *Assign cost to tasks*
- *Play simulation*
- *Add animation to TimeLiner*
- *Export animation from TimeLiner*
- *Link TimeLiner to external scheduling software*

INTRODUCTION

In the previous chapter, you learned about creating viewpoints, cross-sections, and viewpoint animation. In this chapter, you will learn about the concept of Timeliner, creating tasks, preparing schedules, assigning cost, attaching objects to tasks, and creating 4D simulation.

INTRODUCTION TO TIMELINER

The Timeliner feature in Navisworks is used to visualize the construction sequence of the model before it begins. To plan a construction sequence, you need the schedules that include duration of various tasks necessary to complete the work. Using this schedule, you can estimate the total cost and resources allocated to each task. Thus, for visualizing the construction sequence of a model, you need to create a 4D simulation. To create a 4D simulation, you need a 3D model and the construction schedule. In 4D simulation, the planning sequence of construction activities will be displayed with respect to time. While playing the 4D simulation, you will be able to see the effects of schedule on the model and can compare planned dates against actual dates. This comparison helps you assess the delay in project completion.

The benefits of 4D are as follows:
1) The ability to understand the full scope of the project
2) Improved the project execution strategy
3) Enhanced communication and team building
4) The ability to identify time and space conflicts

THE TIMELINER WINDOW

The **TimeLiner** window is a dockable window that can be used to add tasks, attach objects in the model to the tasks, set durations, and simulate project schedules, as shown in Figure 5-1.

Tip
The options in the TimeLiner window will be activated only when you open an existing project.

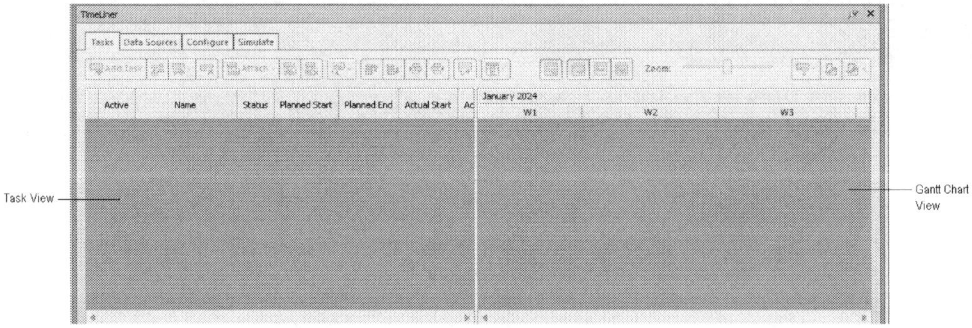

Figure 5-1 *The* ***TimeLiner*** *window*

The **TimeLiner** window consists of four tabs: **Tasks**, **Data Sources**, **Configure**, and **Simulate**. These tabs are discussed next.

Tasks Tab

You can use the **Tasks** tab to create and manage various tasks/activities involved in a project. The **Tasks** tab contains two panes: the Task View pane and the Gantt Chart View pane, refer to Figure 5-1.

Data Sources Tab

In the **Data Sources** tab, you can import schedules from other project management softwares such as Microsoft Project, Primavera, and Asta Powerproject. When you choose the **Data Sources** tab, it will be empty. Once you import the schedules, the sources of the added schedules will be displayed in a tabular format in this tab. In this table, the **Name** column displays the name of the new data source. The **Source** column displays the source of the added data such as the name of the source software. The **Project** column displays the name and location of the added data.

Configure Tab

In the **Configure** tab, you can set up the task parameters such as task types and appearance of tasks. This tab displays a list of all the task types in a tabular format, as shown in Figure 5-2.

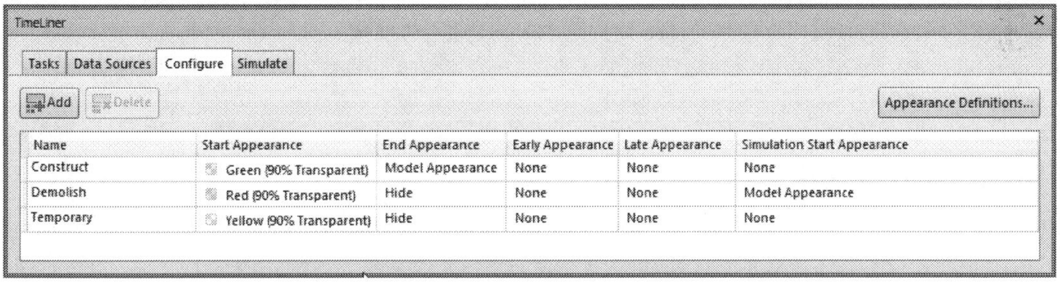

*Figure 5-2 The **Configure** tab in the **TimeLiner** window*

Simulate Tab

In this tab, you will create 4D simulation which illustrates the sequence of construction activities throughout the duration of the project schedule.

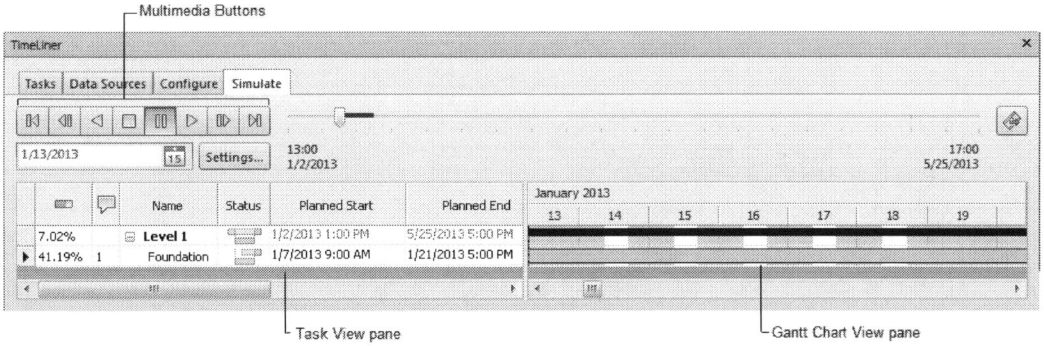

*Figure 5-3 The **Simulate** tab in the **TimeLiner** window*

PLAYING SIMULATION

In Navisworks, simulation is a real-world process of visualizing the construction sequence of a project. After developing the model and linking the tasks with the construction elements, you can play the simulation. Before playing a simulation, ensure that all tasks are active in the **Tasks** tab of the **TimeLiner** window.

EXPORTING ANIMATION FROM THE TIMELINER

You can export the simulation as an AVI file or a sequence of images. To do so, choose the **Export Animation** button on the right of the position slider; the **Animation Export** dialog box will be displayed.

ADDING ANIMATION TO THE TIMELINER

In the **TimeLiner** window, you can add the viewpoint animation and the camera animation to the individual tasks or to the entire schedule. The animation will be played during the simulation as it increases the quality of simulation.

Adding Animation to the Entire Schedule

You can add a viewpoint animation to the entire schedule displayed in the **Tasks** tab of the **TimeLiner** window.

Adding Animation to Tasks

In Navisworks, you can also add animation to an individual task.

TIMELINER COSTS

There are five cost types available in **TimeLiner**: **Material Cost**, **Labor Cost**, **Equipment Cost** **Subcontractor Cost**, and **Total Cost**. These cost types are used to track the cost of a project throughout the schedule.

IMPORTING SCHEDULES FROM EXTERNAL SCHEDULING SOFTWARE

In the **TimeLiner** window, you can import schedules from other software.

TUTORIALS

General instructions for downloading tutorial files:

1. Download the *c05_navisworks_2024_tut zip file* for this tutorial from *https://www.cadsofttech.com*. The path of the file is as follows: ***Books > Civil/GIS Books > Autodesk Navisworks > Exploring Autodesk Navisworks 2024 for Novices (Learn By Doing) > Tutorial Files.***

2. Now, save and extract the downloaded folder at the following location: *C:\ nws_2024\ c05_nws_2024_tut.*

Note

The default unit system used in the tutorials is metric. To change the units to imperial, select the required units from ***Options Editor > Interface > Display Units****.*

Tutorial 1 Creating 4D Simulation

In this tutorial, you will open the *c05_navisworks_2024_tut1.nwf* file and create 4D simulation by using various options available in the **TimeLiner** window. The schedule used in this tutorial has been created in Microsoft Project Professional 2010. **(Expected time: 45 min)**

The following steps are required to complete this tutorial:

a. Open the file *c05_navisworks_2024_tut1.nwf.*
b. Display the **TimeLiner** window.
c. Import the Microsoft project file.
d. Attach model objects to the tasks.
e. Assign task types.
f. Specify the simulation parameters.
g. Play the simulation.
h. Save the project.

Opening the File

In this section, you will open the file to be used in this tutorial.

1. Choose the **Open** button from the Quick Access Toolbar; the **Open** dialog box is displayed.

2. In this dialog box, browse to the following location: *C:\nws_2024\c05_nws_2024_tut.*

3. Select **Navisworks File Set (*.nwf)** from the **Files of type** drop-down list.

4. Select the file **c05_navisworks_2024_tut1** from the displayed list of files; the file name is displayed in the **File name** edit box.

5. Next, choose the **Open** button in the dialog box; the model is displayed in the Scene View, as shown in Figure 5-4.

Figure 5-4 Model displayed in the Scene View

Displaying the TimeLiner Window

In this section, you will display the **TimeLiner** window.

1. Choose the **TimeLiner** tool from the **Tools** panel of the **Home** tab; the **TimeLiner** window is displayed, as shown in Figure 5-5.

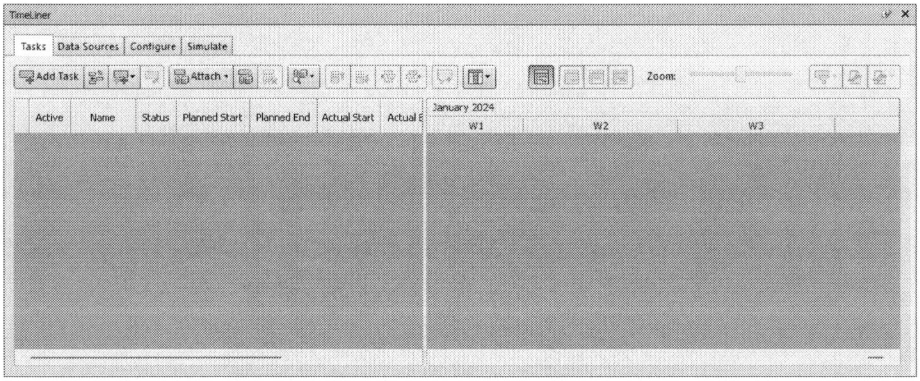

Figure 5-5 *The* ***TimeLiner*** *window*

Importing the Schedule Created in Microsoft Project Professional

In this section, you will import the construction schedule created in the Microsoft Project software.

1. In the **TimeLiner** window, choose the **Data Sources** tab.

2. In the **Data Sources** tab, choose the **Add** button; a drop-down list is displayed.

3. Select the **Microsoft Project 2007-2013** option from the drop-down list; the **Open** dialog box is displayed.

4. In this dialog box, browse to the following location: *C:\nws_2024\c05_nws_2024_tut.*

5. Select **Construction_Project** from the dialog box, as shown in Figure 5-6. Now, choose the **Open** button; the **Field Selector** dialog box is displayed.

Note

You need to install the Microsoft Project 2016 or any other version on the system to perform this tutorial.

6. In the **Field Selector** dialog box, click in the **External Field Name** column corresponding to **Synchronization ID**; a drop-down list is displayed, refer to Figure 5-5.

7. Next, select the **ID** option from the drop-down list, as shown in Figure 5-7

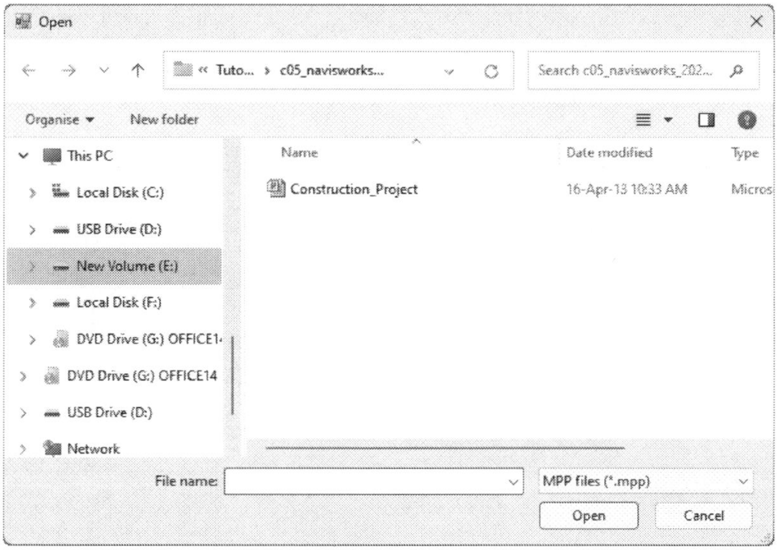

Figure 5-6 *Selecting the **Construction_Project** option from the **Open** dialog box*

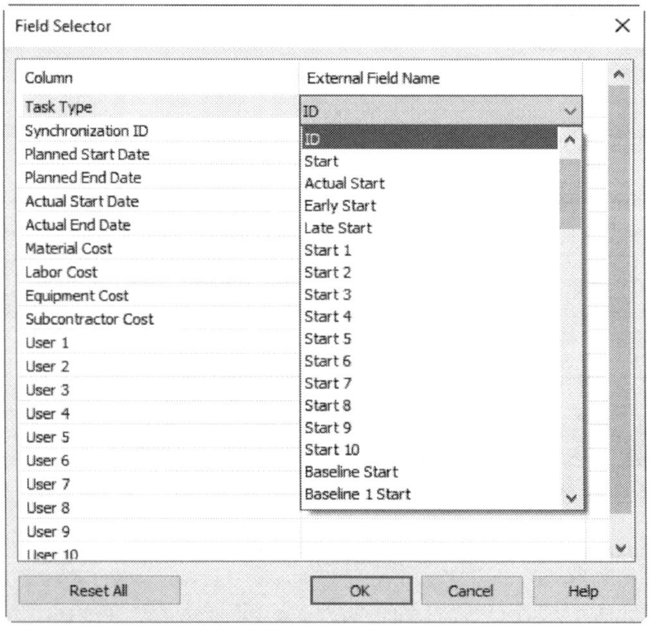

Figure 5-7 *Selecting the* ***ID*** *option for* ***Synchronization ID***

8. Similarly, specify the options for **Planned Start Date**, **Planned End Date**, **Actual Start Date**, **Actual End Date**, and **Material Cost**, refer to Figure 5-8.

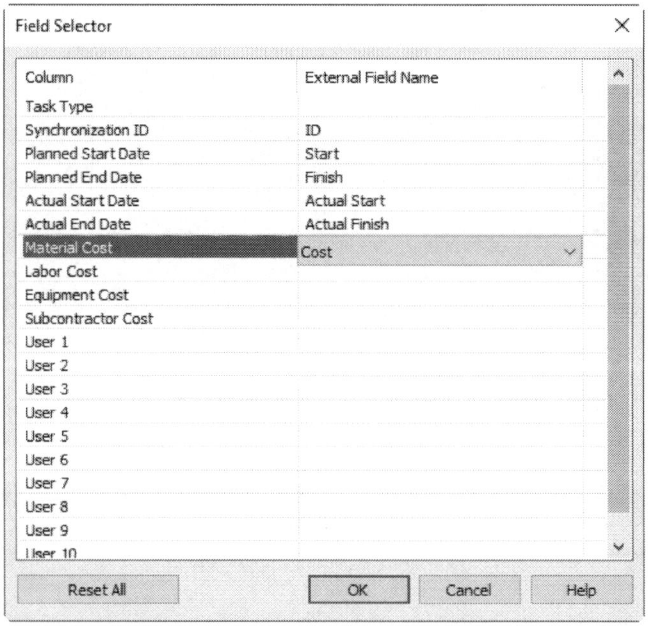

Figure 5-8 *Specifying options in the* ***Field Selector*** *dialog box*

9. Next, choose the **OK** button; the dialog box is closed and the new data source is added in the **Data Sources** tab of the **TimeLiner** window, refer to Figure 5-9.

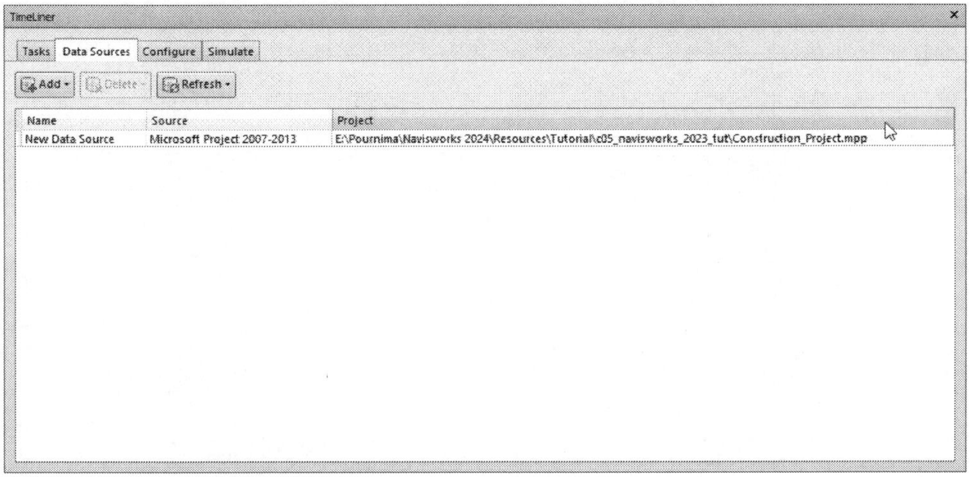

Figure 5-9 New Data Source *added in the* ***Data Sources*** *tab*

10. Next, select the **New Data Source** text under the **Name** column and right-click on it; a shortcut menu is displayed.

11. Choose the **Rebuild Task Hierarchy** option from the menu, as shown in Figure 5-10.

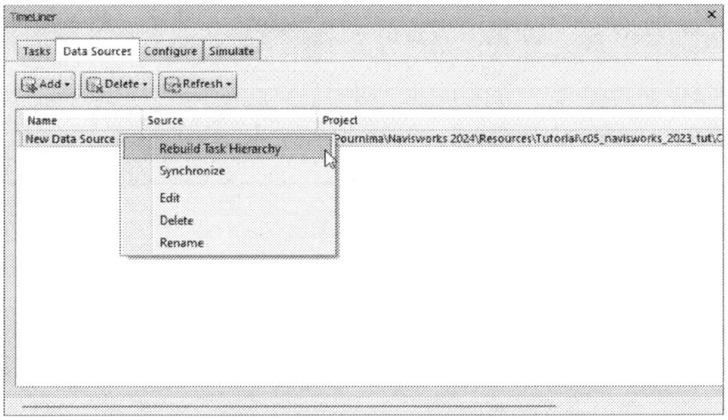

Figure 5-10 *Choosing the* ***Rebuild Task Hierarchy*** *option from the shortcut menu*

12. Next, choose the **Tasks** tab; the schedule is displayed in the **Tasks** tab, refer to Figure 5-11.

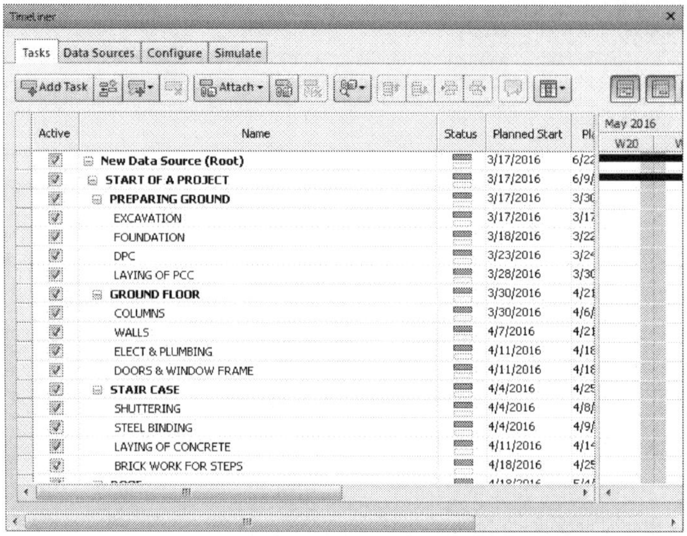

*Figure 5-11 Schedule displayed in the **Tasks** tab*

Attaching Model Objects to the Tasks

In this section, you will attach the model objects to the tasks.

1. Select the **Manage Sets** option from the **Sets** drop-down list in the **Select & Search** panel of the **Home** tab; the **Sets** window is displayed, as shown in Figure 5-12.

*Figure 5-12 The **Sets** window*

2. Select **Structural Foundations** in the **Sets** window; the selected objects are highlighted in the Scene View.

3. Choose the **Hide Unselected** button from the **Visibility** panel of the **Home** tab; the unselected objects are hidden in the Scene View and only the selected foundations are displayed, as shown in Figure 5-13.

Figure 5-13 *Selected foundations displayed in the Scene View*

4. In the **TimeLiner** window, select the **FOUNDATION** task under the **Name** column in the **Tasks** tab.

5. Now, select the **Attach Current Selection** option from the **Attach** drop-down list, refer to Figure 5-14; the selected structural foundations are attached to the **FOUNDATION** task.

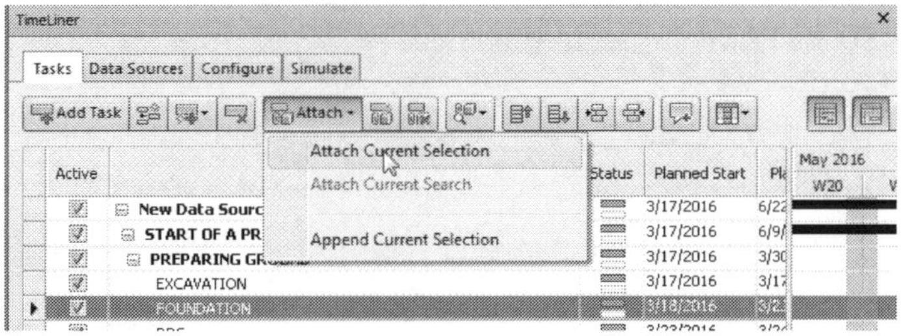

Figure 5-14 *Selecting the* ***Attach Current Selection*** *option from the list*

6. Next, choose the **Hide Unselected** button from the **Visibility** panel of the **Home** tab; the unselected objects become visible in the Scene View.

7. Select **Floor Slab** in the **Sets** window.

8. Select **DPC** and press the CTRL key, and then select the **LAYING OF PCC** task in the **TimeLiner** window.

9. Next, select the **Attach Current Selection** option from the **Attach** drop-down list; the selected objects are attached to the tasks.

10. Now, select **Columns** in the **Sets** window.

11. In the **TimeLiner** window, under the **Ground Floor** head, select the **Columns** sub task and then select the **Attach Current Selection** option from the drop-down list; the selected columns are attached to the task.

12. Next, select **Walls** in the **Sets** window.

13. Select the **Walls** sub task under the **Ground Floor** head in the **TimeLiner** window and then select the **Attach Current Selection** option from the drop-down list; the selected walls are attached to the task.

14. Select **Stairs** in the **Sets** window.

15. Select the **STAIR CASE** task in the **TimeLiner** window and then select the **Attach Current Selection** option from the drop-down list; the stairs are attached to the task.

16. Select **Beams** in the **Sets** window.

17. Attach the selected beams to the **BEAM** task under the **ROOF** head in the **TimeLiner** window.

18. Select **Roof Slab** in the **Sets** window; the roof slab is highlighted in the Scene View.

19. Next, attach the selected object to the **SLAB** task under the **ROOF** head in the **TimeLiner** window and select the **Attach Current Selection** option from the drop-down list; the roof slab is attached to the task.

20. Select **Top roof beam** in the **Sets** window.

21. Attach the selected beams to the **BEAM** task under the **TOP ROOF** head in the **TimeLiner** window.

22. Next, select **Top roof** in the **Sets** window.

23. Attach the selected objects to the **SLAB** task under the **TOP ROOF** head in the **TimeLiner** window.

24. Select **Railings** in the **Sets** window; the selected objects are highlighted in the Scene View.

25. Attach the selected railing to the **RAILING** task in the **TimeLiner** window.

26. Select **Boundary Walls** in the **Sets** window and assign it to **BOUNDARY WALL** task in the **TimeLiner** window.

27. Select **Doors** and **Windows** in the **Sets** window and assign them to **DOORS** and **WINDOWS GLASS** task, respectively under the **FINISHING** head in the **TimeLiner** window.

Assigning Task Types

In this section, you will assign the task type to the required tasks.

1. In the **TimeLiner** window, click in the **Task Type** column corresponding to the **FOUNDATION** task in the **Tasks** tab; a drop-down list will be displayed.

Note

*To make the **Task Type** column visible in the **TimeLiner** window, it is recommended to expand the table displayed in it.*

2. Select the **Construct** option from the list.

3. Repeat the procedure followed in steps 1 and 2 and assign the **Construct** task type to the following tasks: **DPC**, **LAYING OF PCC**, **COLUMNS**, **WALLS**, **STAIR CASE**, **BEAM** and **SLAB**; **BEAM** and **SLAB** under the **ROOF** task; **BEAM**, **SLAB**, **RAILING**, **BOUNDARY WALL** under the **TOP ROOF** task; **DOORS**, and **WINDOW GLASS** tasks under the **FINISHING** task.

Specifying the Simulation Parameters

In this section, you will specify the simulation parameters.

1. Choose the **Configure** tab in the **TimeLiner** window and ensure that **Green (90% Transparent)** is specified in the **Start Appearance** column corresponding to the **Construct** task type.

2. Next, choose the **Settings** button in the **Simulate** tab; the **Simulation Settings** dialog box is displayed.

3. Specify the playback duration as **20seconds** in the **Playback Duration** edit box, refer to Figure 5-15.

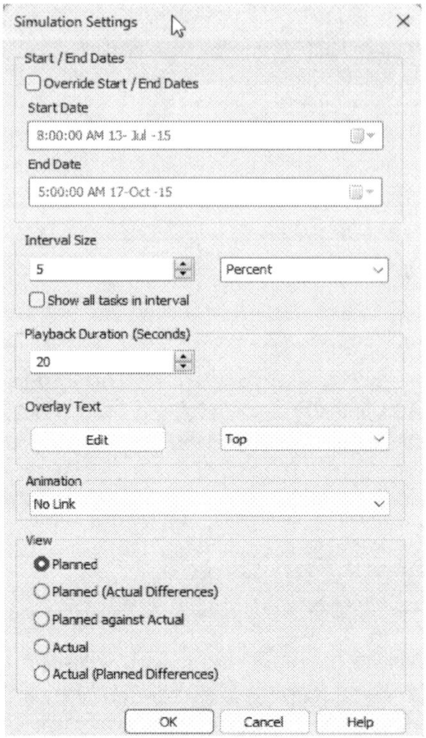

Figure 5-15 *The* ***Simulation Settings*** *dialog box*

4. In the **Start/End Dates** area, select the **Override Start / End Dates** check box; the **Start Date** and **End Date** edit boxes are enabled.

5. In the **Start Date** edit box, click on the down arrow; a calendar is displayed.

6. Select the date 13th July, 2015 from the flyout; the date is displayed in the edit box.

7. Repeat the procedure followed in steps 5 and 6 and specify the date 17th October 2015 in the **End Date** edit box.

8. Specify the time **08:00:00 AM** and **5:00:00 PM** in the **Start Date** and **End Date** edit boxes, respectively, refer to Figure 5-15.

9. Next, select the **Actual** radio button in the **View** area.

10. Next, choose the **Edit** button in the **Overlay Text** area; the **Overlay Text** dialog box is displayed, refer to Figure 5-16.

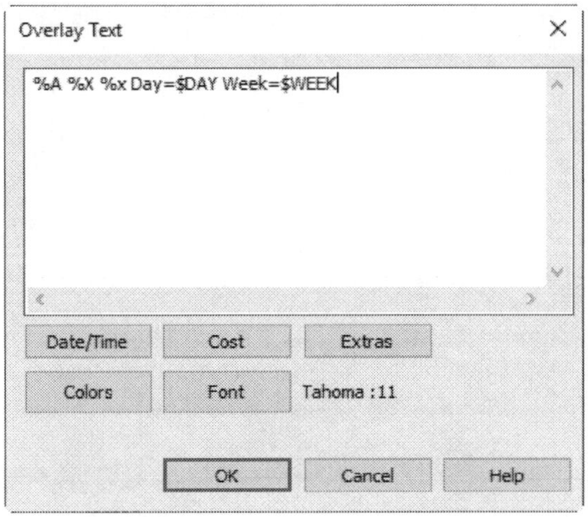

Figure 5-16 *The* ***Overlay Text*** *dialog box*

11. Delete the text displayed in the text box.

12. Choose the **Date/Time** button; a flyout is displayed. Select the **Date and time representation appropriate for locale** option from the flyout.

13. Next, choose the **Extras** button; a menu is displayed. Select the **Currently active tasks** and **Number of days since start** options from the menu.

14. Choose the **Cost** button; a menu is displayed. Select **Total Cost** from the menu.

15. Choose the **OK** button in the **Overlay Text** dialog box; the dialog box is closed.

16. Next, select the **Show all tasks in interval** check box in the **Interval Size** area of the **Simulation Settings** dialog box.

17. Choose the **OK** button; the **Simulation Settings** dialog box is closed.

Playing the Simulation

In this section, you will play the simulation of construction sequence.

1. Close the **Sets** window.

2. Now, choose the **Play** button available at the top in the **Simulate** tab; the simulation starts playing.

Saving the Project

In this section, you will save a project.

1. To save the project with the current view, choose **Save As** from the Application Menu, the **Save As** dialog box is displayed.

2. Browse to the *Navisworks_2024* folder and enter **c05_navisworks_2024_tut01** in the **File name** edit box.

3. Choose the **Navisworks File Set (*.nwf)** file format from the **Save as type** drop-down list, and then choose the **Save** button.

Tutorial 2 4D Sequence with Crane Animation

In this tutorial, you will open the *c05_navisworks_2024_tut2.nwf* file, and create 4D simulation of a building project, and link an object animation with the sequence.

(Expected time: 30min)

The following steps are required to complete this tutorial:

a. Open the *c05_navisworks_2024_tut2.nwf* file.
b. Create construction schedule in the **TimeLiner** window.
c. Link the crane animation with the task.
d. Specify the simulation parameters.
e. Play the simulation.
f. Save the project.

Opening the File

In this section, you will open the file to be used in this tutorial.

1. Choose the **Open** button from the Quick Access Toolbar; the **Open** dialog box is displayed.

2. In this dialog box, browse to the following location:
   C:\ nws_2024\c05_nws_2024_tut

3. Select **Navisworks File Set (*.nwf)** from the **Files of type** drop-down list; a list of files is displayed.

4. Select the file **c05_navisworks_2024_tut2** from the list; the file name appears in the **File name** edit box.

5. Next, choose the **Open** button in the dialog box; the model is displayed in the Scene View.

Creating The Construction Schedule

In this section, you will create the construction schedule based on the objects saved as selection sets. Here, the tasks will be created using the Auto-Add Tasks method.

1. Choose the **TimeLiner** tool from the **Tools** panel of the **Home** tab; the **TimeLiner** window is displayed in the Scene View. 

2. In the **Tasks** tab of the window, choose the **Auto-Add Tasks** button; a drop-down list is displayed.

3. Select the **For Every Set** option from this list; the tasks are created in the Task View area.

After creating the tasks, you need to arrange them in a construction order and assign the start and end dates of each task.

4. Select the task **Columns** in the Task View area, and click the **Move Up** button twice to move this task up in the list. 

5. Repeat the procedure followed in step 4 and arrange the remaining tasks, refer to Figure 5-17.

6. Specify the start and end dates of each task, refer to Figure 5-17.

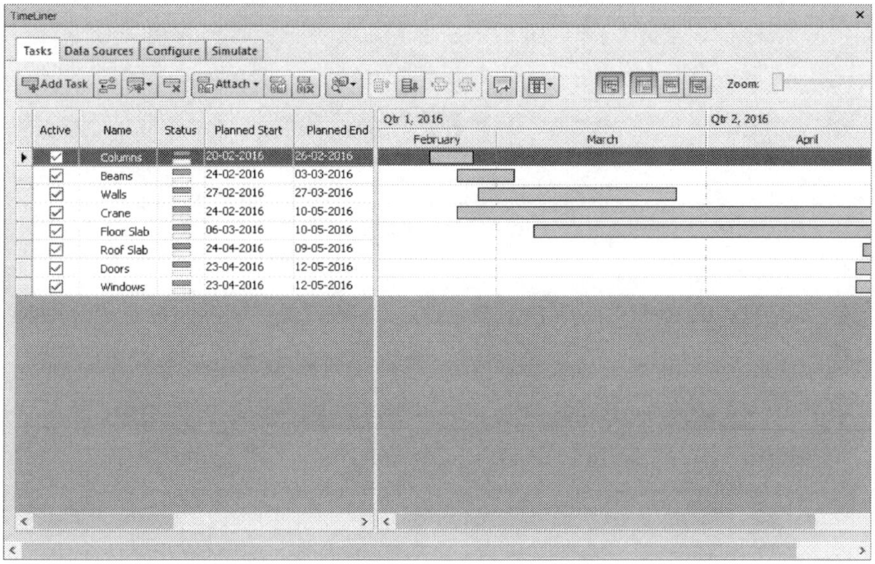

Figure 5-17 *Crane task added in the Task View area of the* ***TimeLiner*** *window*

Linking the Crane Animation with the Crane Task

In this section, you will link the crane animation which has been created using the options available in the **Animator** window.

1. Choose the **Columns** button and select the **Choose Columns** option from the list displayed; the **Choose TimeLiner Columns** dialog box is displayed. 

2. In this dialog box, select the **Animation** and **Animation Behavior** check boxes and choose the **OK** button; the **Animation** and **Animation Behavior** columns are added in the Task View area.

3. Now, select the **Crane** task and click in the **Animation** column; a drop-down list is displayed.

4. Select the **Scene folder\Tower Crane** option from this list and ensure that the **Scale** option is selected in the **Animation Behavior** column.

5. Next, click in the **Task Type** column for the **Crane** task; a drop-down list is displayed.

6. Select the **Temporary** option from the list.

Specifying the Simulation Parameters

In this section, you will specify the simulation parameters.

1. Choose the **Simulate** tab in the **TimeLiner** window and then choose the **Settings** button; the **Simulation Settings** dialog box is displayed.

2. In the **Overlay Text** area of this dialog box, choose the **Edit** button; the **Overlay Text** dialog box is displayed.

3. Choose the **Extras** button; a flyout is displayed. Select the **Number of days since start** option from the flyout.

4. Choose the **Colors** button and select the **Red** option from the flyout. Choose the **OK** button to apply the changes.

5. Retain the other settings in the **Simulation Settings** dialog box and choose the **OK** button to close it.

Playing the Simulation

In this section, you will play the simulation.

1. Choose the **Play** button in the **Simulate** tab of the **TimeLiner** window. Figure 5-18 shows various sequences while playing the simulation along with the animation.

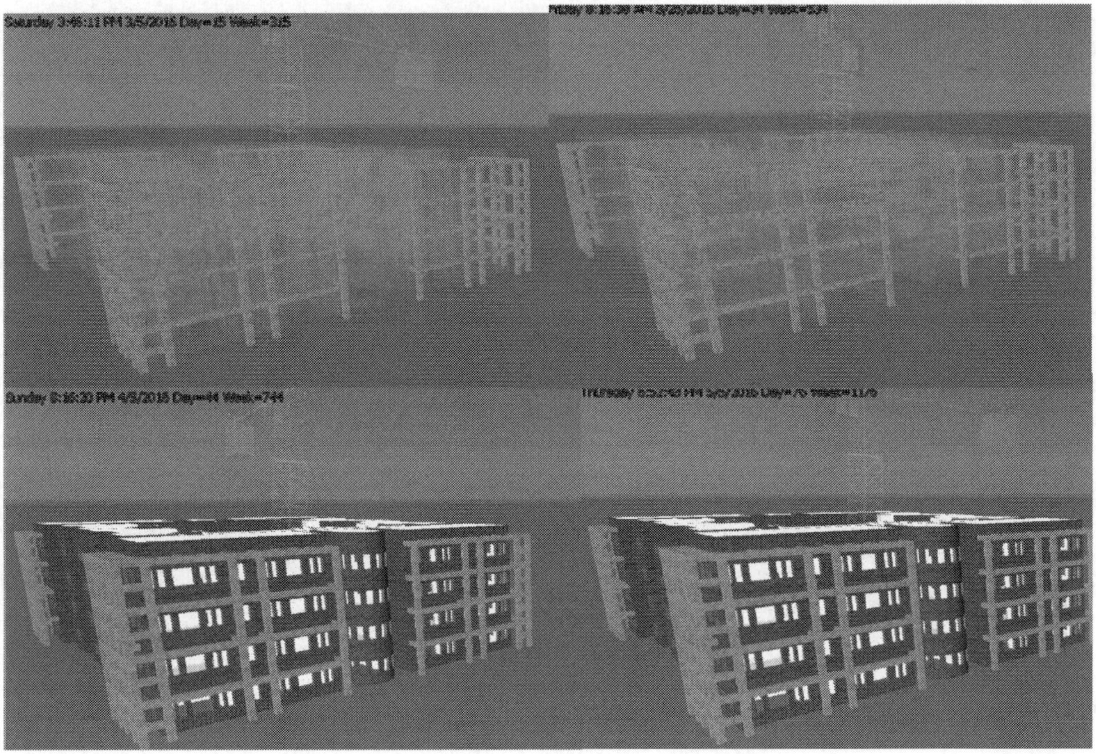

Figure 5-18 *Sequence of the simulation and animation*

Saving the Project

In this section, you will save the project.

1. To save the project with the current view, choose **Save As** from the Application Menu; the **Save As** dialog box is displayed.

2. Browse to the *nws_2024 \ c05_nws_2024_tut* folder and enter **c05_navisworks_2024_tut02** in the **File name** edit box.

3. Now, select the **Navisworks File Set (*.nwf)** file format from the **Save as type** drop-down list, and choose the **Save** button.

SELF-EVALUATION TEST

Answer the following questions and then compare them to those given at the end of this chapter:

1. You can use the __________ option to export the Timeliner schedule in CSV format.

2. The __________ button is used to filter the tasks based on their status.

3. The __________ option is used to rebuild the task hierarchy with the latest data.

4. The __________ option is used to update the existing schedule with the latest data.

5. You can use the __________ task type for the tasks to be constructed during simulation.

6. In Navisworks, you can create selection sets of the objects attached in the Timeliner hierarchy. (T/F)

7. In the **TimeLiner** window, you can add comments to the tasks. (T/F)

8. You can attach same objects to multiple tasks. (T/F)

9. Timeliner rules cannot be imported or exported. (T/F)

10. Comments cannot be added to Timeliner tasks. (T/F)

REVIEW QUESTIONS

Answer the following questions:

1. Which of the following options is used to add tasks in the **TimeLiner** window?

   a) **Refresh** b) **Add Task**
   c) **Attach** d) **Delete**

2. Which of the following tabs contains the options for importing the schedules?

   a) **Simulate** b) **Data Sources**
   c) **Configure** d) **Tasks**

3. Which of the following buttons is used to invoke the **Simulation Settings** dialog box?

   a) **Refresh** b) **Clear Attachment**
   c) **Settings** d) **Simulate**

4. Which of the following options is used to create selection sets of objects attached to the tasks?

a) **Filter by Status** b) **Export To Sets**
c) **Add Comment** d) **Update**

5. Which of the following options is used to define the task as temporary?

a) **Construct** b) **Demolish**
c) **Temporary** d) **Synchronize**

6. Model appearance during simulation cannot be configured. (T/F)

7. You cannot change the color of text displayed during the simulation. (T/F)

8. You cannot link camera animation to the tasks in Timeliner. (T/F)

9. In the Timeliner, you cannot import schedules from other software. (T/F)

10. The **Clear Attachment** button is used to detach the object from the task. (T/F)

EXERCISE

Exercise 1 Creating 4D Simulation

Download and open the *c05_navisworks_2024_ex1* file from *https://www.cadsofttech.com.* Create a 4D simulation by using the options available in the **TimeLiner** window. Figure 5-19 shows the model to be used in this exercise.

(Expected time: 1hr)

The following steps are required to complete this exercise:

a. Open the file *c05_navisworks_2024_ex1.*
b. Import the *Construction_exercise* schedule created in Microsoft Project Professional.
c. Attach model objects to the tasks and specify the simulation parameters.
d. Play the simulation.

e. Save the project as *c05_navisworks_2024_ex01*.

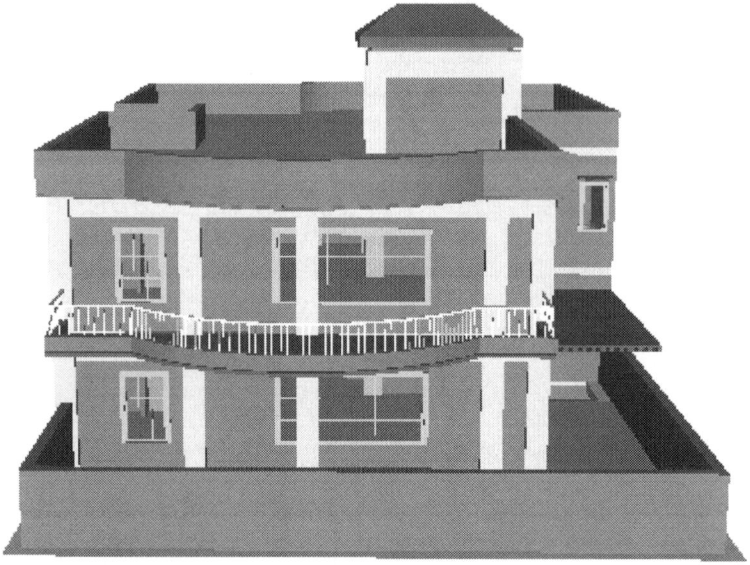

Figure 5-19 *The Residence Building*

Answers to Self-Evaluation Test

1. Export CSV, **2. Filter by Status**, **3. Rebuild Task Hierarchy**, **4. Synchronize**, **5. Construct**,

Chapter 6

Working with Animator and Scripter

Learning Objectives

After completing this chapter, you will be able to:

- *Use the Animator window*
- *Use the Scripter window*
- *Enable scripts*

INTRODUCTION

In the previous chapter, you learned about saving viewpoints, creating viewpoint animation, and creating cross-sections of the model. In this chapter, you will learn to create object animation and adding scripts to the animation by using the options available in the **Animator** and **Scripter** windows. These two windows comprise several options that are used for creating object animation. The options in these windows are discussed in detail in this chapter.

THE ANIMATOR WINDOW

In Navisworks, the **Animator** window is provided to create object animation by capturing every movement of the given object. The options available in the **Animator** window allow you to create several types of animations, animation viewpoint. as shown in Figure 6-1.

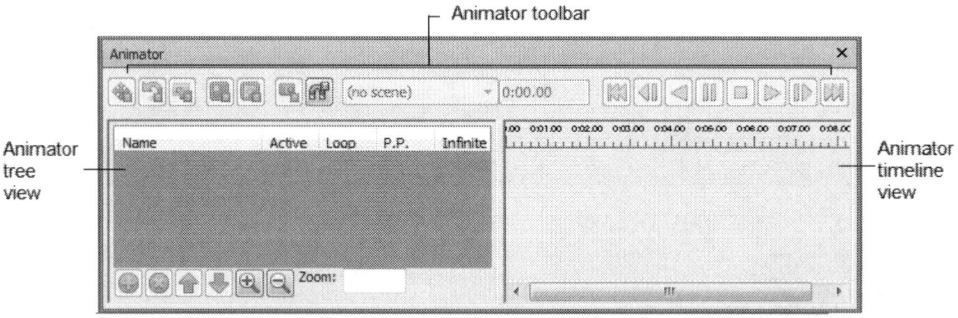

*Figure 6-1 The **Animator** window*

Animator Toolbar

The Animator toolbar is available at the top in the **Animator** window. It contains buttons and options used while creating object animation, as shown in Figure 6-2.

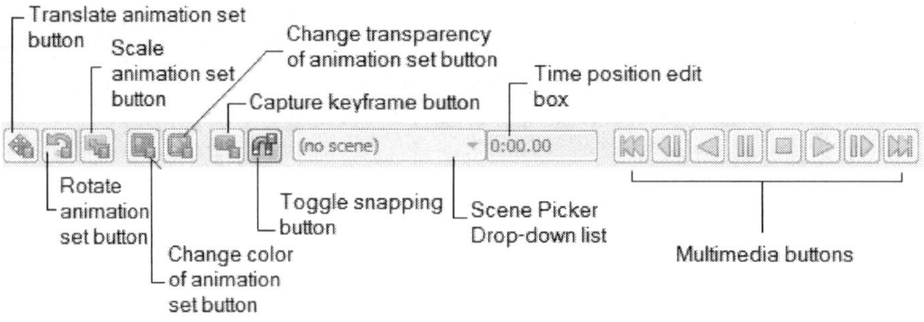

*Figure 6-2 Buttons and options in the **Animator** toolbar*

Animator Tree View

The Animator tree view is available in the left pane of the **Animator** window, refer to Figure 6-1. The Animator tree view contains a list of all the added scenes and scene components such as animation set, camera, and section planes. These scenes and scene components can be arranged in a hierarchy and grouped in a scene folder. The scenes and scene folders are created by using the buttons available at the bottom of the Animator tree view.

Animator Timeline View

The right pane of the **Animator** window displays the Animator timeline view which refers to timeline with keyframes. It displays the length of overall animation in a bar chart view, as shown in Figure 6-3.

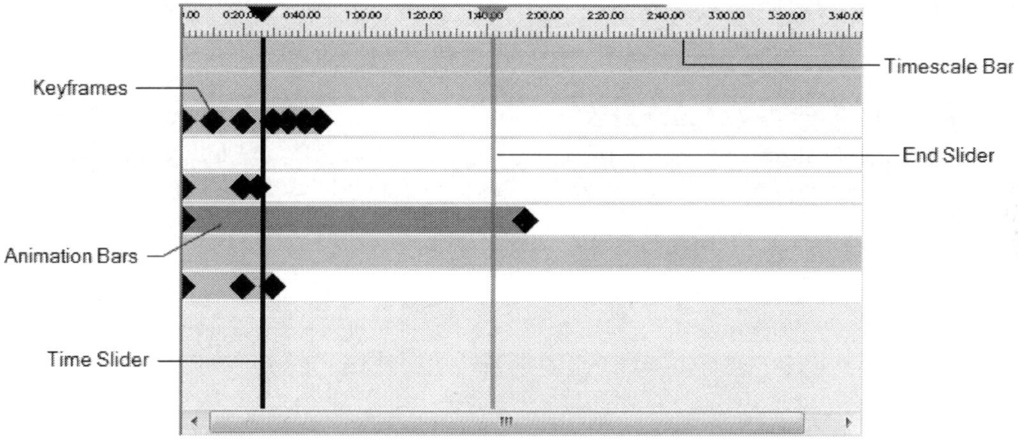

***Figure 6-3** The timeline view in the **Animator** window*

CREATING OBJECT ANIMATIONS

The options available in the **Animator** window are used for creating the object animations. Before creating an animation, you need to create a scene folder which will contain all the scene components.

Creating Object Animation Using the Animation Set

The animation set is created on the basis of currently selected objects in the Scene View. You can also create animation set by using the selection/search set previously created in the model.

Creating Animation Using Camera

The object animation using Camera is created by using the existing viewpoints animation or by creating viewpoints using navigation tools. In a scene, either you can add a blank camera and then create the viewpoints or can copy an existing viewpoints animation to the camera. You can add only one camera in a scene in the Animator tree view.

Creating Animation Using the Section Plane

You can also animate the cross-section views of an object by using the **Add Section Plane** option. In this case, a section plane set is created which contains a list of all the cross-sectional cuts of the model and the keyframes.

EDITING KEYFRAMES

After creating animation, you can modify the keyframes to change the position, time, field of view, and so on. The keyframes can be modified based on the method of creating animation: animation sets, camera, and section plane sets.

The methods of editing keyframes are Listed Below :

1) Editing Animation Set Keyframes.
2) Editing Camera Keyframes.
3) Editing Section Plane Keyframes.

THE SCRIPTER WINDOW

Using the **Scripter** window, you can interact with the objects in the model. To interact with the objects in the model, you need to define script. A script is a collection of events and actions that are expected when certain event conditions are met.

The Scripts Area

The **Scripts** area in the **Scripter** window contains a list of all the added scripts in a hierarchical view. In this area, you can create and manage scripts by using the buttons.

The Events Area

The **Events** area contains a list of events associated with the selected script. In this area, you can create, manage, and test events. The buttons in this area are used to define and manage the events. These buttons are available at the bottom of the **Events** area, as shown in Figure 6-4.

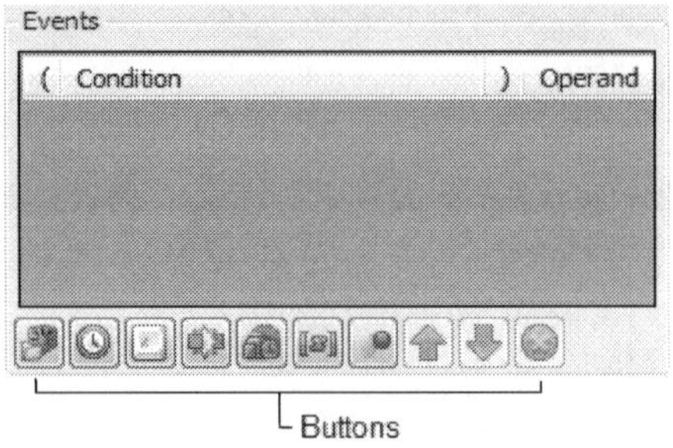

***Figure 6-4** The **Events** area in the **Scripter** window*

The Actions Area

The **Actions** area contains a list of actions associated with the selected script. An action is an activity which is carried out when a script is activated by an event. An action can be playing animation such as opening and closing of the doors. After creating events, you need to define actions for these events. You can add multiple actions to a script. An action will start when the event conditions are met. Using the options in the **Actions** area, you can create, manage, and test actions. The buttons used to define actions are available at the bottom in the **Actions** area, as shown in Figure 6-5.

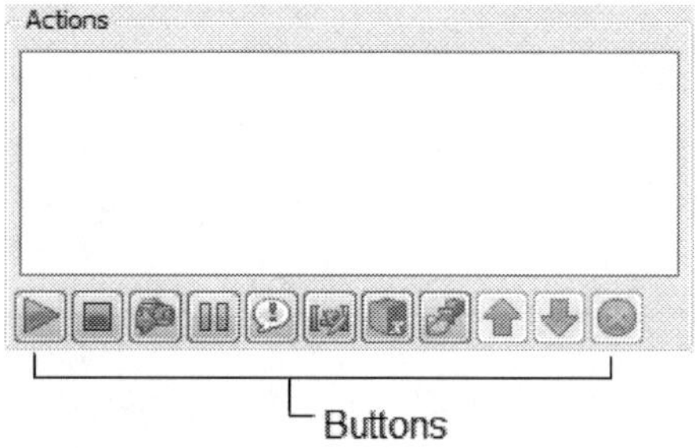

Figure 6-5 *The buttons in the **Actions** area*

The Properties Area

The right pane of the **Scripter** window is the **Properties** area. When you select an event or action, its properties will be displayed in the **Properties** area. These event and action properties can be configured in the **Properties** area.

ENABLING SCRIPTS

After creating the scripts you need to activate them. To do so, choose the **Enable Scripts** button from the **Script** panel in the **Animation** tab; the **Scripter** window will be disabled and now you can interact with the model. To disable the added scripts, choose the **Enable Scripts** button again; the **Scripter** window will be enabled.

TUTORIALS

General instructions for downloading tutorial files:

1. Download the *c06_navisworks_2024_tut* zip file for this tutorial from *https://www.cadsofttech.com*.The path of the file is as follows: ***Books > Civil/GIS Books > Autodesk Navisworks > Exploring Autodesk Navisworks 2024 for Novices (Learn By Doing).***

2. Now, save and extract the downloaded folder at the following location:
C:\ nws_2024\ c06_nws_2024_tut

Note
The default unit system used in the tutorials is metric. To change the units to imperial, select the required units from ***Options Editor > Interface > Display Units****.*

Tutorial 1 Creating Animation

In this tutorial, you will open the *c06_nws_2024_tut1.nwf* file and create object animation using the options in the **Animator** window. **(Expected time: 45min)**

The following steps are required to complete this tutorial:

a. Open the *c06_nws_2024_tut1.nwf* file.
b. Display the **Saved Viewpoints** window.
c. Display the **Animator** window.
d. Create a scene folder.
e. Add a scene in the scene folder.
f. Create animation using Animation set.
g. Play animation.
h. Save the project.

Opening the Existing Model

In this section, you will open the model *c06_nws_2024_tut1* that was created using the Revit software.

1. Choose the **Open** button from the Quick Access Toolbar; the **Open** dialog box is displayed on the screen.

2. In this dialog box, browse to the following location:
C:\nws_2024\c06_nws_2024_tut

3. Select the **Navisworks File Set (*.nwf)** file format from the **Files of type** drop-down list.

4. Next, select *c06_nws_2024_tut1* from the displayed list of files; the file is displayed in the **File name** edit box.

5. Now, choose the **Open** button; the model is displayed in the Scene View, as shown in Figure 6-6.

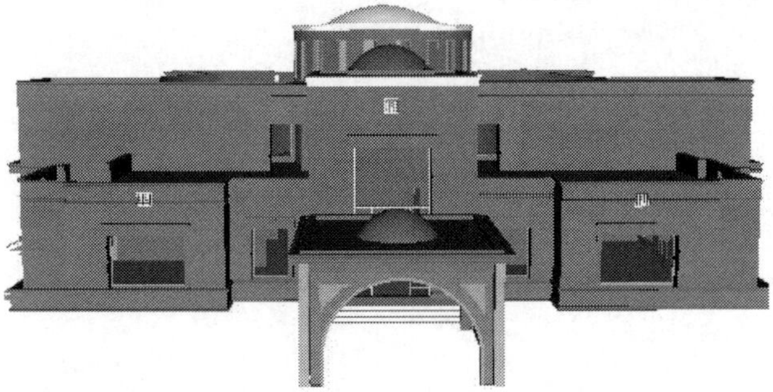

Figure 6-6 Model displayed in the Scene View

Displaying the Saved Viewpoints Window

In this section, you will display the **Saved Viewpoints** window and display the view.

1. Choose the **Saved Viewpoints Dialog Launcher** tool from **Viewpoint > Save, Load & Playback > Save Viewpoint drop-down**; the **Saved Viewpoints** window is displayed, as shown in Figure 6-7.

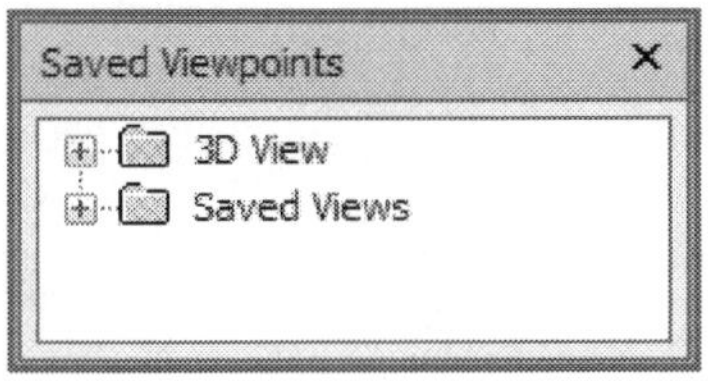

Figure 6-7 The ***Saved Viewpoints*** *window*

2. In this window, expand the **Saved Views** folder and select the **Door** viewpoint; the door is displayed in the Scene View, as shown in Figure 6-8. Next, close the **Saved Viewpoints** window.

Figure 6-8 The door displayed in the scene view

Displaying the Animator Window

In this section, you will display the **Animator** window.

1. Choose the **Animator** tool from the **Tools** panel in the **Home** tab; the **Animator** window is displayed in the Scene View.

Creating a Scene Folder

In this section, you will create a scene folder and group various scene components in this folder.

1. Click the **Add Scene** button at the lower left corner of the **Animator** window; a flyout is displayed, as shown in Figure 6-9.

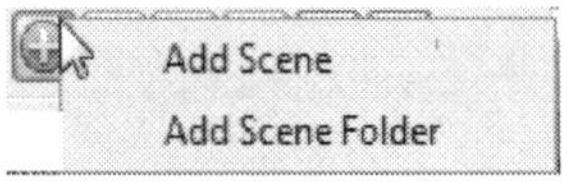

Figure 6-9 *The flyout displayed on clicking the* ***Add Scene*** *button*

Note

The options in the ***Animator*** *window will not be enabled if the* ***Enable Scripts*** *button is chosen from the* ***Scripts*** *panel in the* ***Animation*** *tab.*

2. Choose the **Add Scene Folder** option from the flyout; a scene folder is created in the **Animator** tree view. Note that the name of this folder is in editable mode.

3. Specify the name **Animation Set** in the edit box, as shown in Figure 6-10, and press ENTER.

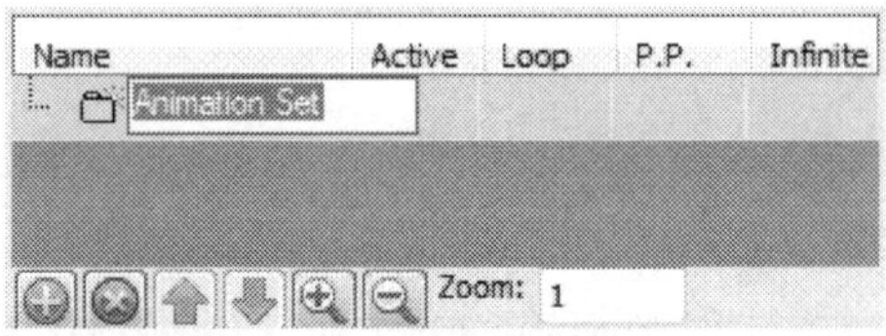

Figure 6-10 *The scene folder*

Creating a Scene in the Scene Folder

In this section, you will create a scene in the **Animation Set** folder.

1. Choose the **Add Scene** button at the bottom of the **Animator** window; a flyout is displayed.

2. Choose **Add Scene** from the flyout; a scene with the name **Scene 1** is added to the **Animation Set** folder.

3. Next, double-click on **Scene 1**; an edit box is displayed.

4. Specify the name **Door** in the edit box, refer to Figure 6-11 and press ENTER.

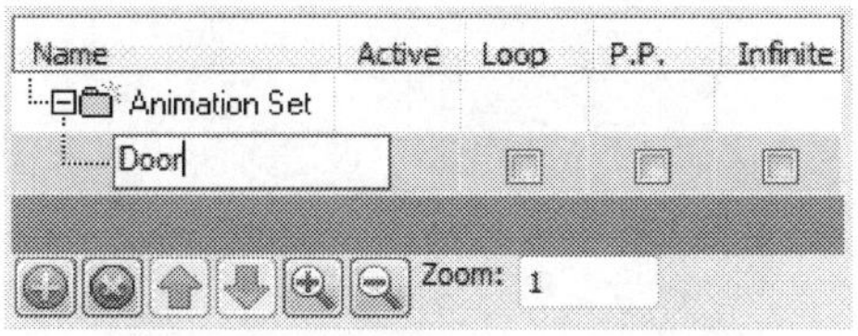

*Figure 6-11 The scene added in the **Animation Set** folder*

Note
You can also create scene folders and their components using the options in the shortcut menu displayed on right-clicking in the tree view.

Creating Animation Using the Animation Set

In this section, you will create animations in the **Animation Set** scene by using the **Translate animation set** and **Rotate animation set** tools.

1. Invoke the **Select** tool from the Quick Access Toolbar and select the door and the door frame in the Scene View by using the CTRL key, as shown in Figure 6-12.

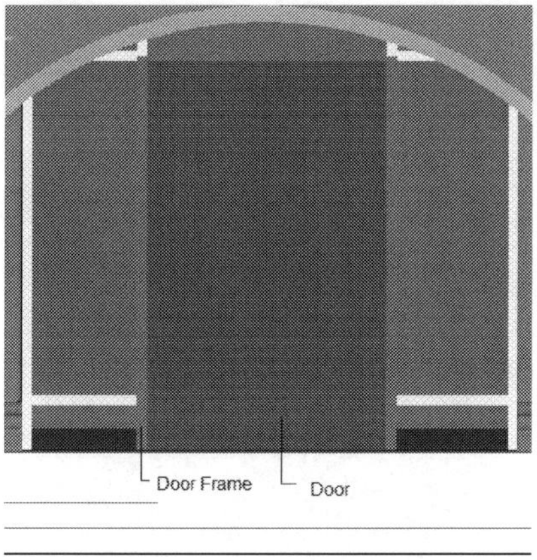

Figure 6-12 The door and door frame selected in the model

2. Next, choose the **Hide Unselected** button from the **Visibility** panel in the **Home** tab; the unselected objects are hidden and the selected objects are visible, refer to Figure 6-13.

Note
*On choosing the **Hide Unselected** button if the **Autodesk Navisworks Manage 2024** message box is displayed then choose the **OK** button.*

Figure 6-13 *Selected objects in the Scene View*

3. Deselect the selected objects by clicking in the Scene View and select only the door.

4. Using the **Orbit** tool, rotate the door, as shown in Figure 6-14.

Figure 6-14 *The rotated door in the Scene View*

5. Next, right-click on the **Door** in the Animator tree view of the **Animator** window; a shortcut menu is displayed.

6. In this menu, choose **Add Animation Set > From current selection**, refer to Figure 6-15; **Animation Set 1** is added under **Door**, as shown in Figure 6-16.

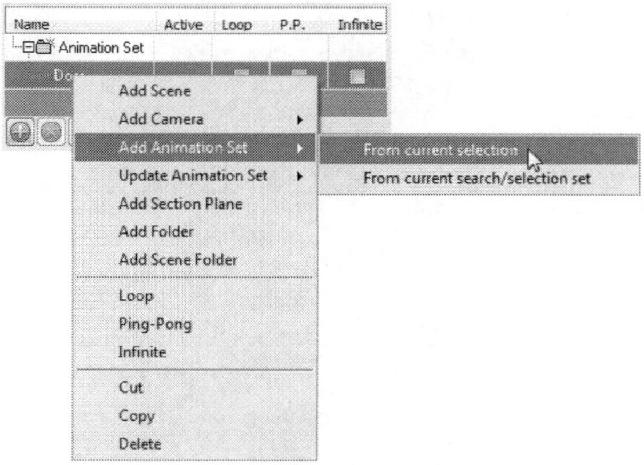

Figure 6-15 *Choosing an option from the shortcut menu*

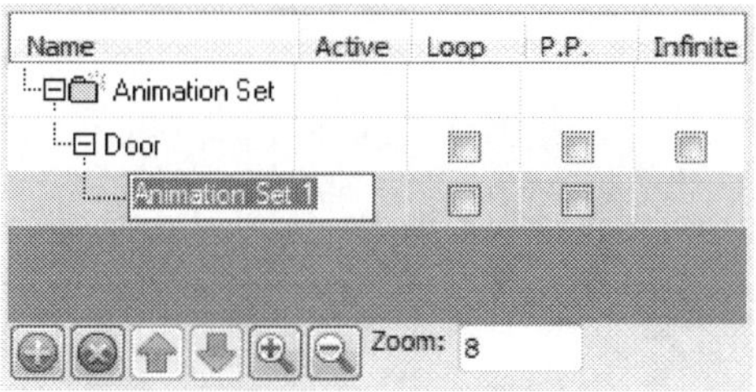

Figure 6-16 *The Animation Set created in the Animator tree view*

7. Rename **Animation Set 1** as **Opening Door** and press ENTER.

8. Choose the **Capture keyframe** button in the Animator toolbar to capture the initial position of the model.

9. Specify the time as **0:05.00** in the **Time position** edit box.

10. Choose the **Rotate Animation Set** button from the Animator toolbar; the **Rotate** gizmo appears in the Scene View, as shown in Figure 6-17.

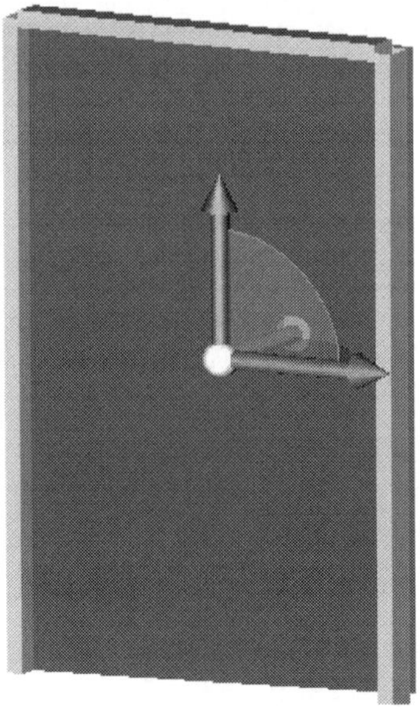

Figure 6-17 *The **Rotate** gizmo displayed in the Scene View*

11. Place the cursor on the center ball of the gizmo; the ball is highlighted and a hand cursor appears on the screen, as shown in Figure 6-18.

12. Press and hold the left mouse button and then drag the gizmo and place it, as shown in Figure 6-19.

Figure 6-18 *Cursor at the center ball of the Gizmo*

Figure 6-19 *Gizmo placed at the lower corner*

13. Place the cursor on the green curve; the curve is highlighted and a hand cursor appears, as shown in Figure 6-20.

14. Press and hold the left mouse button keeping the CTRL key pressed and rotate the gizmo, as shown in Figure 6-21. Note that you need to rotate the gizmo not the door.

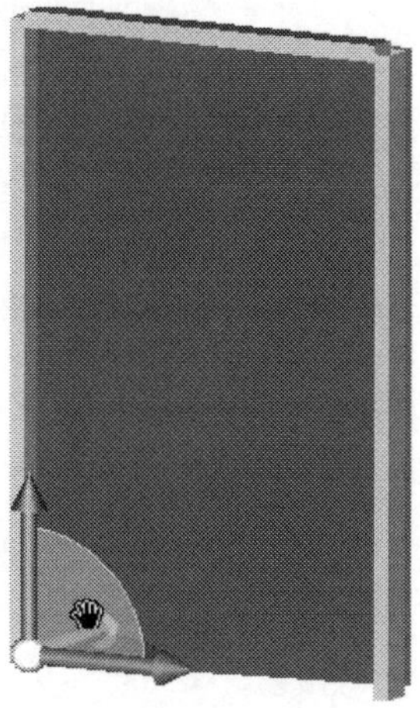

Figure 6-20 *Placing cursor on the green curve*

Figure 6-21 *Gizmo after being rotated*

15. Place the cursor on the blue curve; the curve is highlighted, as shown in Figure 6-22.

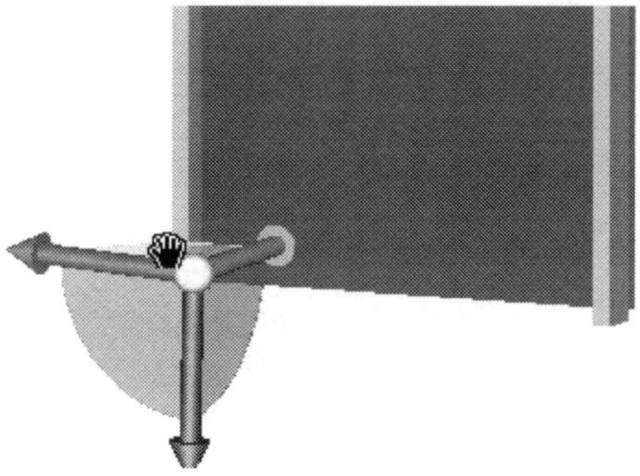

Figure 6-22 *Cursor placed on the blue curve*

16. Press and hold the left mouse button and drag the gizmo slowly in the left direction to open the door, as shown in Figure 6-23. After opening the door, release the left mouse button.

17. Choose the **Rotate animation set** button in the Animator toolbar; the **Rotate** gizmo disappears.

18. Next, choose the **Capture keyframe** button to capture the current position of the door.

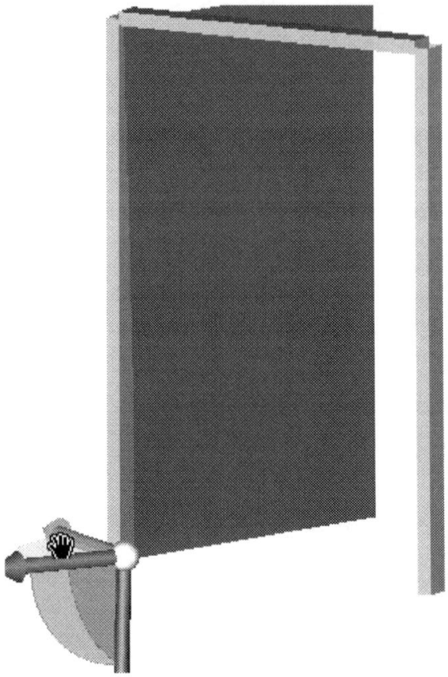

Figure 6-23 *Gizmo Moved in left direction to open door*

19. Create another animation set, as shown in Figure 6-24, following the procedure explained in the earlier section.

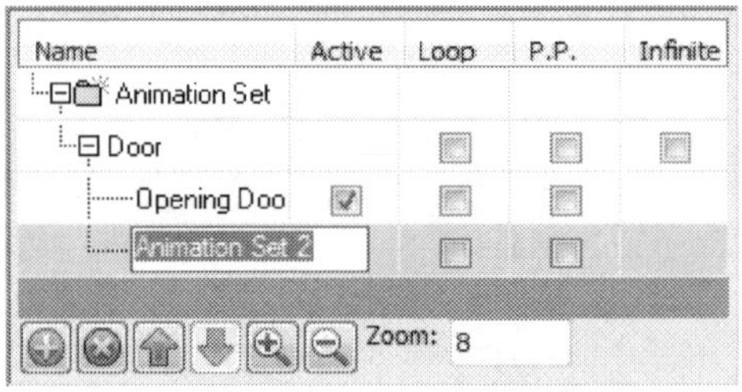

Figure 6-24 *Creating another Animation Set*

20. Rename the Animation Set as **Closing Door**.

21. Specify the time as **0:08.00** in the **Time position** edit box and press ENTER.

22. Choose the **Capture Keyframe** button to capture the current position of the selected door.

23. Specify the time as **0:13.00** in the **Time position** edit box.

24. Next, choose the **Rotate animation set** tool from the Animator toolbar; the **Rotate** gizmo is displayed in the Scene View. Place the gizmo, as shown in Figure 6-25.

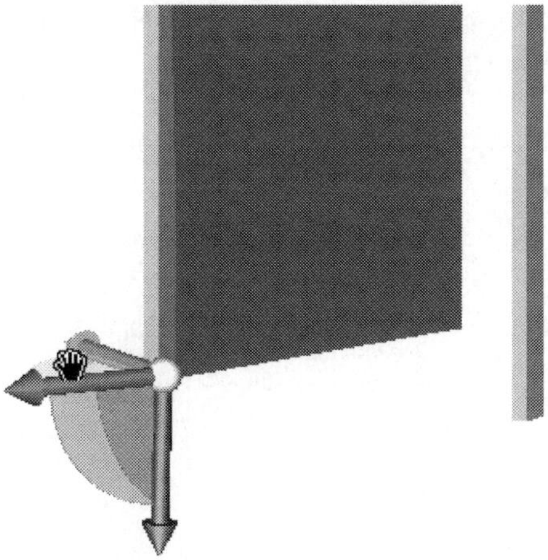

Figure 6-25 *The **Rotate** gizmo displayed in the Scene View*

25. Place the cursor on the blue curve of the gizmo; the curve is highlighted and a hand cursor appears.

26. Press and hold the left mouse button and then drag the mouse in the left direction to close the door, refer to Figure 6-26.

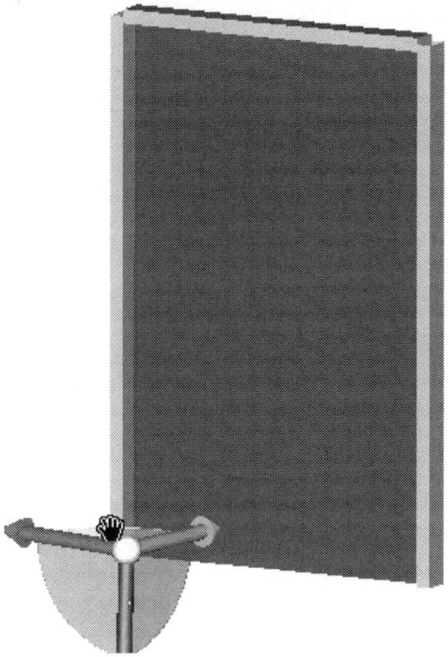

Figure 6-26 *The closed door*

27. Next, choose the **Rotate animation set** tool from the **Animator** toolbar; the **Rotate** gizmo disappears.

28. Choose the **Capture keyframe** button to capture the current position of the door.

29. Next, choose the **Unhide All** option from **Home > Visibility > Unhide All** drop-down; all the objects become visible in the Scene View.

30. Next, deselect the selected objects and display the **Door** viewpoint under the **Saved Views** folder in the **Saved Viewpoints** window.

Playing Animation

In this section, you will play the animation.

1. Select the **Door** animation set in the Animator tree view of the **Animator** window and choose the **Play** button from the Animator toolbar to bring back the animation to the start and play both the animations.

Saving the Project

In this section, you will save the project.

1. Close the **Animator** window and choose **Save As** from the Application Menu; the **Save As** dialog box is displayed.

2. Browse to the *nws_2024/c06_nws_2024_tut* folder and enter **c06_nws_2024_tut01** in the **File name** edit box. Choose the **Navisworks File Set (*.nwf)** file format from the **Save as type** drop-down list, and then choose the **Save** button.

Tutorial 2 Creating Scripts

In this tutorial, you will open the *c06_nws_2024_tut2.nwf* file and create scripts in the **Scripter** window. **(Expected time: 45min)**

The following steps are required to complete this tutorial:

a. Open the *c06_nws_2024_tut2.nwf* file.
b. Display the **Scripter** window.
c. Create the script folder.
d. Add script in the script folder.
e. Add events.
f. Add actions.
g. Adjust the model
h. Enable scripts.
i. Navigate inside the model.
j. Save the project.

Opening the Model

In this section, you will open the model *c06_nws_2024_tut2* which was created in Revit software.

1. To open the existing model, choose the **Open** button from Quick Access Toolbar; the **Open** dialog box is displayed on the screen.

2. In this dialog box, browse to the following location:
   C:\nws_2024\c06_nws_2024_tut

3. Select **Navisworks File Set (*.nwf)** file format from the **Files of type** drop-down list.

4. Next, select the *c06_nws_2024_tut2* from the displayed list of files; the file is displayed in the **File name** edit box.

5. Now choose the **Open** button; the model is displayed in the Scene View.

Displaying the Scripter Window

In this section, you will display the **Scripter** window.

1. Choose the **Scripter** tool from the **Tools** panel in the **Home** tab; the **Scripter** window is displayed in the Scene View.

Creating the Script Folder

In this section, you will create script folder in the **Scripter** window.

1. Choose the **Add New Folder** button available at the lower left corner in the **Scripter** window; a new folder is created. Note that the name of this folder is displayed in editable mode, refer to Figure 6-27.

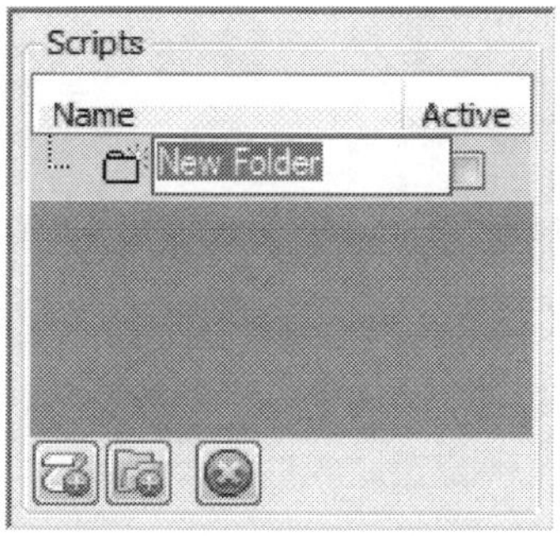

***Figure 6-27** Specifying the folder name in the **Scripts** area*

2. Specify the name **Door_Scripts** in the edit box and press ENTER.

Creating Scripts in the Script Folder

In this section, you will add scripts in the script folder.

1. Choose the **Add New Script** button available at the lower left corner in the **Scripter** window; a new script is added to the **Door_Scripts** folder and is displayed in an edit box, refer to Figure 6-28.

*Figure 6-28 New Script added in the **Door_Scripts** folder*

2. Rename the scripts as **Open & Close Door** and press ENTER.

Adding Events and Configuring Event Properties

In this section, you will add events and configure event properties.

1. Choose the **On Key Press** button available at the bottom of the **Events** area; the **On Key Press** event is added to the **Events** area, as shown in Figure 6-29.

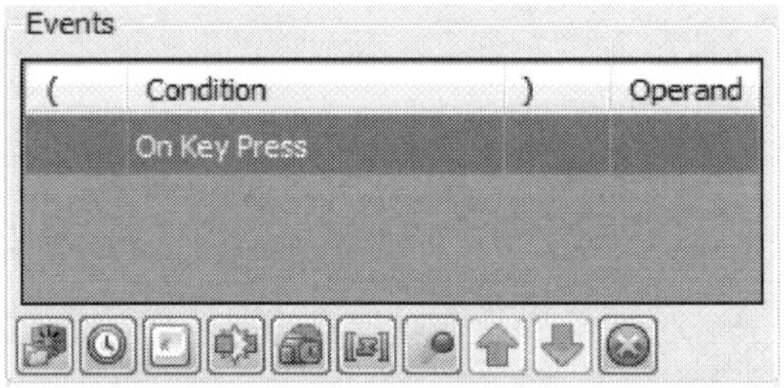

*Figure 6-29 The **On Key Press** event added to the **Events** area*

Note

*On adding the event, its properties are displayed in the **Properties** area, as shown in Figure 6-30.*

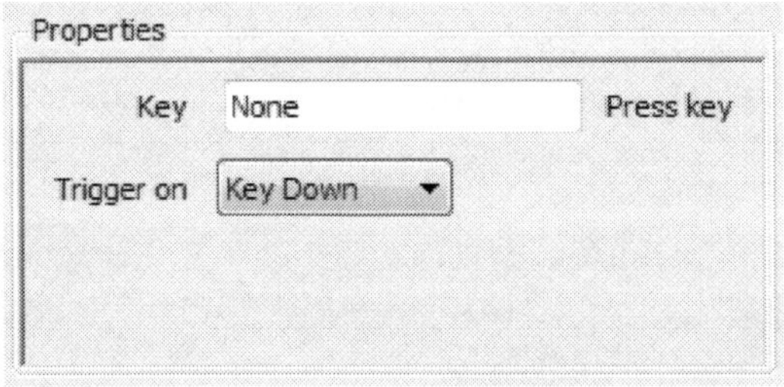

*Figure 6-30 The **On Key Press** event properties displayed to the **Properties** area*

2. Place the cursor and click in the **Key** edit box in the **Properties** area, refer to Figure 6-30.

3. Next, press the **Up** key from the keyboard; the **Up** key is specified in the **Key** edit box in the **Properties** area, refer to Figure 6-31.

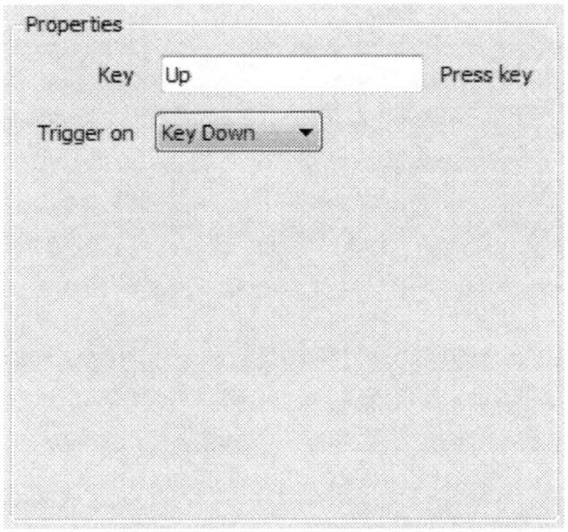

Figure 6-31 *The **Up** key specified in the **Key** edit box*

4. Select the **Key Up** option from the **Trigger on** drop-down list in the **Properties** area, refer to Figure 6-32.

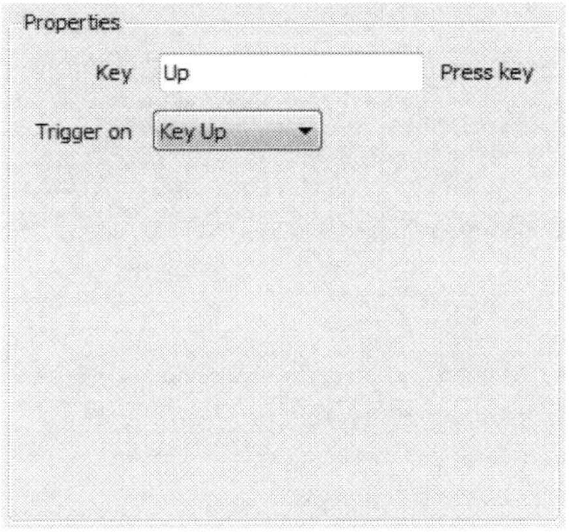

Figure 6-32 *The **Key Up** option selected from the **Trigger on** drop-down list*

5. Choose the **On Animation** button  available at the bottom in the **Events** area; the **On Animation** event is added to the **Events** area, as shown in Figure 6-33.

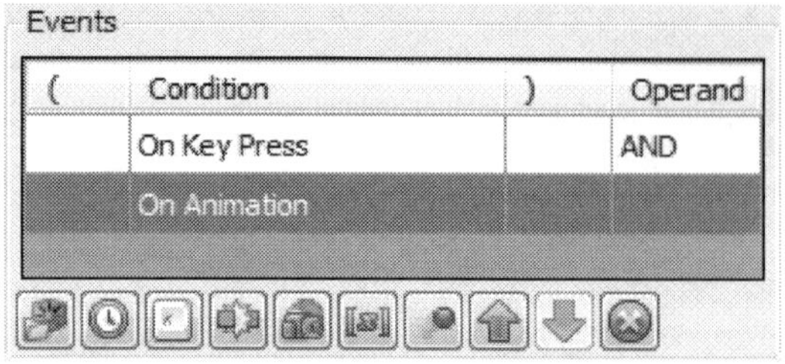

Figure 6-33 *The* ***On Animation*** *event added to the* ***Events*** *area*

6. Click in the **Operand** column corresponding to the **On Key Press** event in the **Events** area; a drop-down list is displayed, as shown in Figure 6-34.

7. Select the **OR** option from the displayed drop-down list.

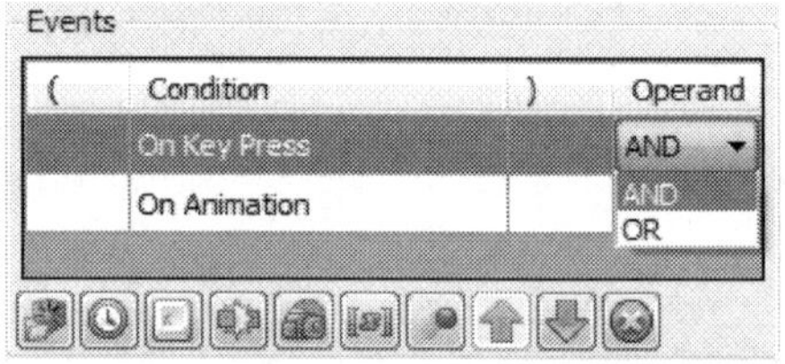

Figure 6-34 *The drop-down list displayed in the* ***Operand*** *column*

8. Select the **On Animation** event in the **Events** area; its properties are displayed in the **Properties** area, as shown in Figure 6-35.

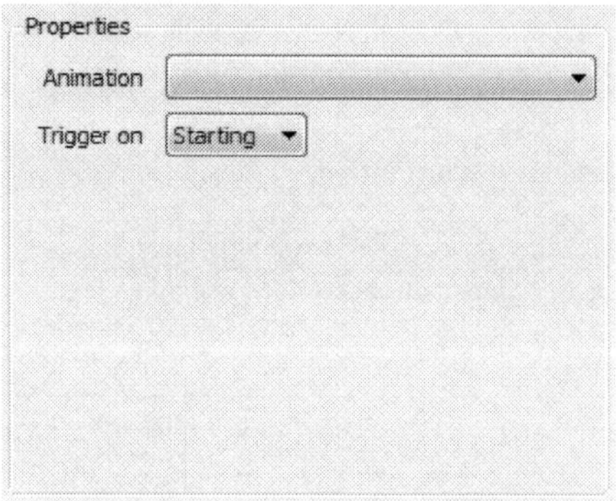

Figure 6-35 *The* ***On Animation*** *event properties displayed in the* ***Properties*** *area*

9. Select the **Opening Door** animation from the **Animation** drop-down list in the **Properties** area, refer to Figure 6-36.

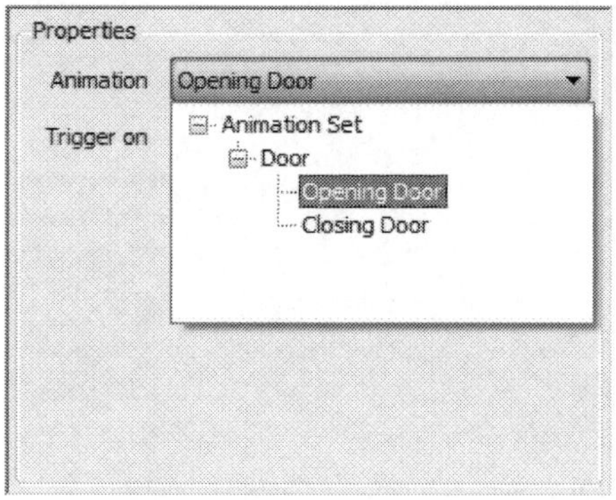

***Figure 6-36** Selecting the **Opening Door** option from the **Animation** drop-down list*

10. Select the **Ending** option from the **Trigger on** drop-down list, as shown in Figure 6-37.

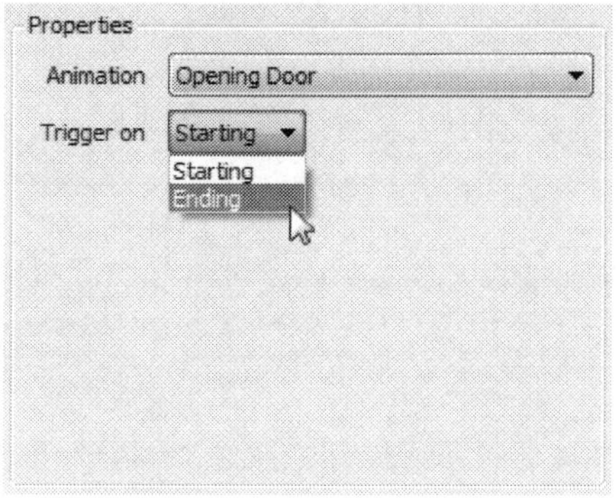

***Figure 6-37** Selecting the **Ending** option from the **Trigger on** drop-down list*

Adding Actions and Configuring Their Properties

In this section, you will add actions and configure their properties.

1. Choose the **Play Animation** button  available at the bottom of the **Actions** area ; the **Play Animation** action is added to the **Actions** area, refer to Figure 6-38. Also, the properties of the **Play Animation** are displayed in the **Properties** area, as shown in Figure 6-39.

***Figure 6-38** The **Play Animation** action added to the **Actions** area*

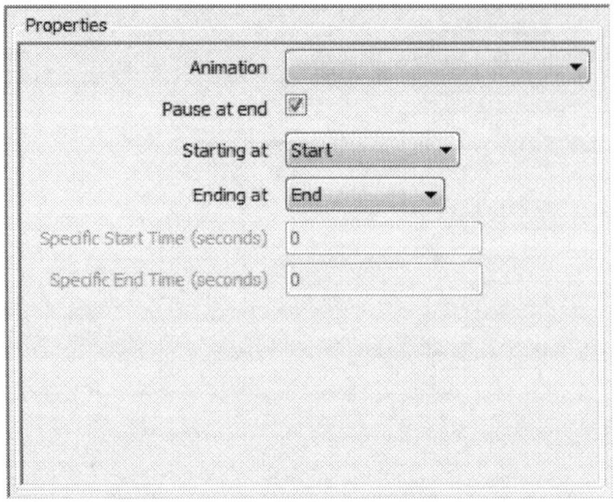

***Figure 6-39** The **Play Animation** properties displayed in the **Properties** area*

2. Select the **Opening Door** option from the **Animation** drop-down list in the **Properties** area, refer to Figure 6-40.

3. Ensure that the **Start** and **End** options are selected in the **Starting at** and **Ending at** drop-down lists. Also, make sure the **Pause at end** check box is selected, refer to Figure 6-41.

4. Repeat step 1 and add the **Play Animation** action again to the **Actions** area, refer to Figure 6-42. Its properties are displayed in the **Properties** area, as shown in Figure 6-43.

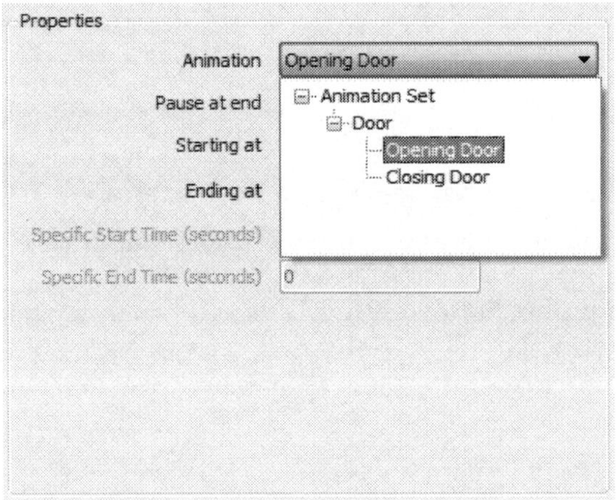

***Figure 6-40** Selecting the **Opening Door** animation from the **Animation** drop-down list*

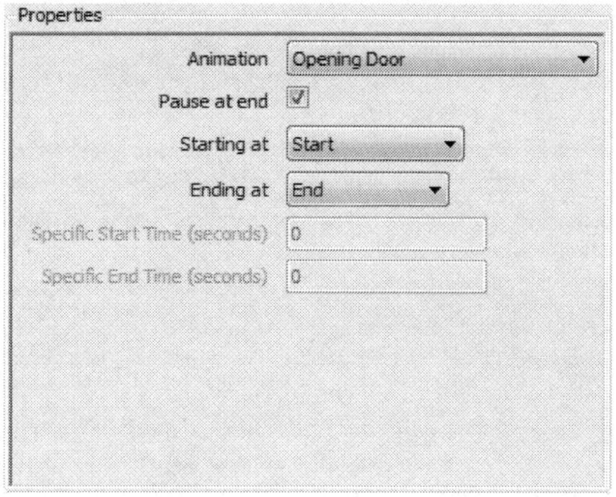

Figure 6-41 *The* ***Start*** *and* ***End*** *options selected in corresponding drop-down lists*

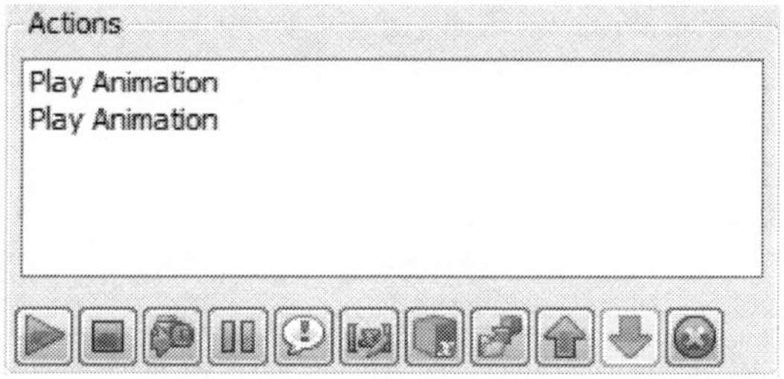

Figure 6-42 *The* ***Play Animation*** *actions added in the* ***Actions*** *area*

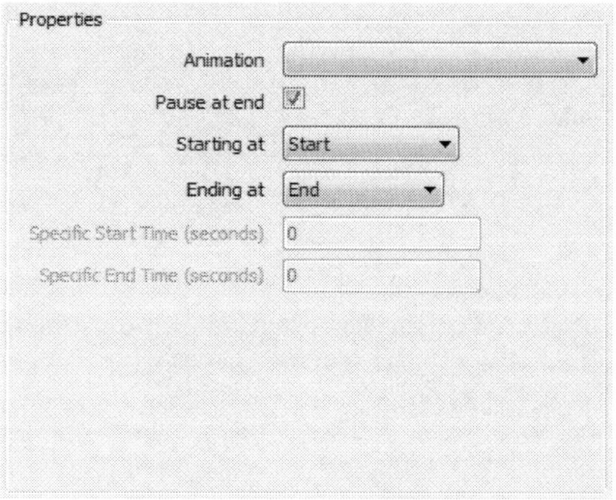

Figure 6-43 *The* ***Play Animation*** *properties displayed in the* ***Properties*** *area*

5. Select the **Closing Door** option from the **Animation** drop-down list in the **Properties** area, refer to Figure 6-44.

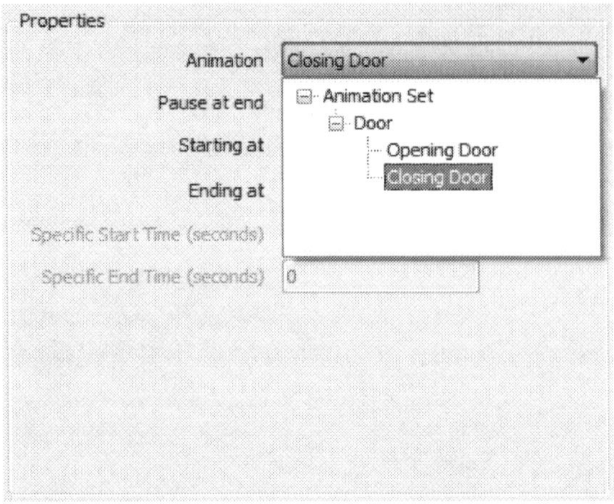

***Figure 6-44** Selecting the **Closing Door** option from the **Animation** drop-down list*

6. Ensure that the **Start** and **End** options are selected in the **Starting at** and **Ending at** drop-down lists, respectively. Also, make sure that the **Pause at end** check box is selected, refer to Figure 6-45.

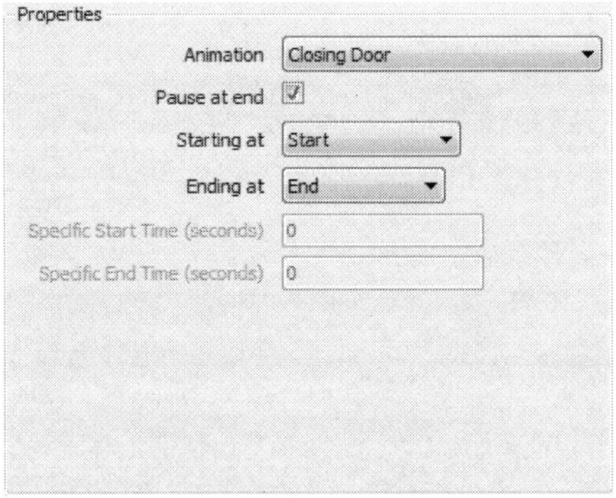

***Figure 6-45** The **Start** and **End** options selected in corresponding drop-down lists*

Tip

*Instead of adding the **On Keypress** and **On Animation** events, you can also add the **On Hotspot** event type. In this case, you will add the event twice and combine them using the **OR** operand. Select the door and then select the **Sphere on selection** option from the **Hotspot** drop-down list and specify the radius value as **0.06** m (0.164 ft) for the first event and **0.50** m (1.64 ft) for the second event. The **Play Animation** action parameters will be same as of the **On Key** press event.*

Adjusting the Model

In this section, you will adjust the model to walk around the model.

1. Choose the **Zoom** tool from the **Zoom** drop-down in the Navigation Bar; the shape of the cursor changes.

2. Press and hold the left mouse button and drag the mouse in the forward direction. Release the mouse button when the model is displayed, as shown in Figure 6-46.

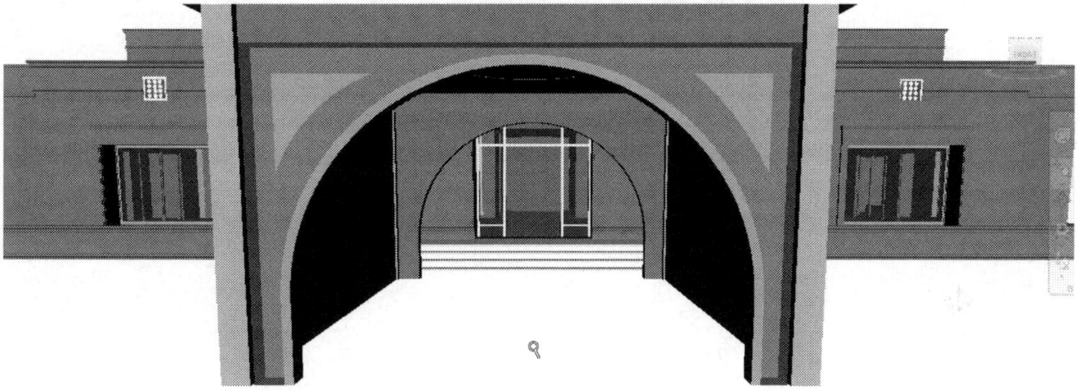

Figure 6-46 *Enlarged view of the model*

Enabling Scripts

In this section, you will enable the scripts defined in the **Scripter** window.

1. Choose the **Enable Scripts** tool from the **Script** panel in the **Animation** tab; the **Scripter** window is deactivated.

Navigating Inside the Model

In this section, you will walk around the model and use the scripts for interaction.

1. Choose the **Walk** tool from the **Walk/Fly** drop-down in the Navigation Bar; the cursor changes into a feet shaped cursor.

2. Press and hold the left mouse button and drag the mouse in forward direction. Release the button when you reach near the door, as shown in Figure 6-47.

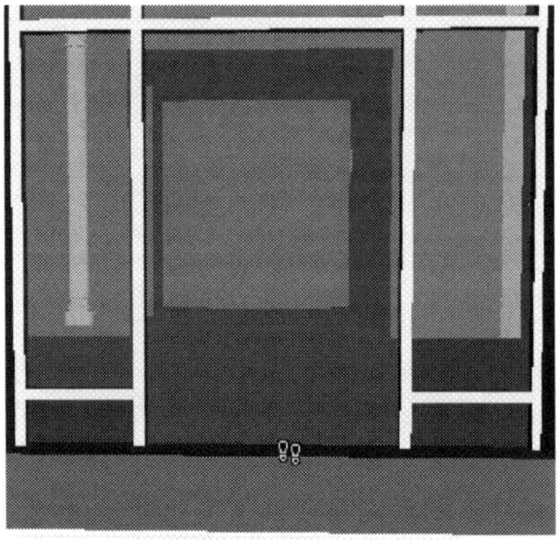

Figure 6-47 *Walking towards the door*

3. Press and release the **Up** arrow key from the keyboard; the door opens, as shown in Figure 6-48.

Figure 6-48 *Door opens after pressing the* ***Up*** *arrow key*

4. Walk inside the model; the **Closing Door** animation starts after completion of the **Opening Door** animation.

5. Using the **Zoom** tool, change the model view, as shown in Figure 6-49.

***Figure 6-49** Model after using the **Zoom** tool*

Tip
*In the **Hotspot** event type, use the avatar while navigating. The parameters for the avatar will be same as the parameters used in the Tutorial 1 of Chapter 2.*

Saving the Project

In this section, you will save the project.

1. To save the project with the current view, choose **Save As** from the Application Menu; the **Save As** dialog box is displayed.

2. Browse to the *nws_2024/c06_nws_2024_tut* folder and enter *c06_nws_2024_tut02* in the **File name** edit box, choose **Navisworks File Set (*.nwf)** file format from the **Save as type** drop-down list and then choose the **Save** button.

SELF-EVALUATION TEST

Answer the following questions and then compare them to those given at the end of this chapter:

1. The __________ dialog box is used to modify the keyframes.

2. The __________ event activates a script using a particular key from the keyboard.

3. The __________ actions will play the animation when a script is triggered.

4. The __________ button in the **Scripter** window is used to add show viewpoint actions.

5. The __________ tool is used to activate scripts.

6. The **On Collision** event activates a script when collision occurs. (T/F)

7. Each scene in the **Animator** window can contain more than one camera animation. (T/F)\

8. You can modify the scripts defined in the **Scripter** window after choosing the **Enable Scripts** button. (T/F)

9. The **Test Logic** option in the **Scripter** window is used to check the logic of an event condition. (T/F)

10. The **Test Action** option in the **Scripter** window is used to verify the execution of an action. (T/F)

REVIEW QUESTIONS

Answer the following questions:

1. Which of the following buttons is used to add an action that closes the current model and loads a new model?

   a) **Play Animation** b) **Store Property**
   c) **Load Model** d) **Set Variable**

2. Which of the following events activate a script when the camera is within a particular range?

   a) **On Timer** b) **On Start**
   c) **On Hotspot** d) **On Animation**

3. Which of the following actions create a text file with a message?

   a) **Set Variable** b) **Pause**
   c) **Send Message** d) **Store Property**

4. Which of the following buttons is used to add the **Pause** action?

   a) **Play** b) **Stop**
   c) **Pause** d) **Load Model**

5. Which of the following buttons is used to add the **On Variable** event?

   a) **Add Scene** b) **Add New Script**
   c) **On Variable** d) **Move**

6. The **Animator** window contains the list of scripts. (T/F)

7. The **Capture keyframe** button is used to play the animation. (T/F)

8. The Time slider bar shows the end position of an animation. (T/F)

9. The **Ping Pong** option is used to play the animation indefinitely. (T/F)

10. In Navisworks, you cannot modify the keyframes. (T/F)

EXERCISE

Exercise 1 Creating Car Animation

Download and open the *c06_nws_2024_ex1* file from *https://www.cadsofttech.com*. Create the vehicle animation using the animation set option in the **Animator** window. Create scripts for stopping the vehicle when they reach the same point. Figure 6-50 shows the model to be used for this exercise. **(Expected time: 1 hr)**

The following steps are required to complete this exercise:

a. Open the *c06_nws_2024_ex1* file.
b. Display the **Saved Viewpoints** window.
c. Display the **Animator** window.
d. Create scene folder.
e. Add scene to the scene folder.
f. Create Animation Set.
g. Create animation.
h. Play animation.
i. Save the project as *c06_naviswork s_2024_ex01*.

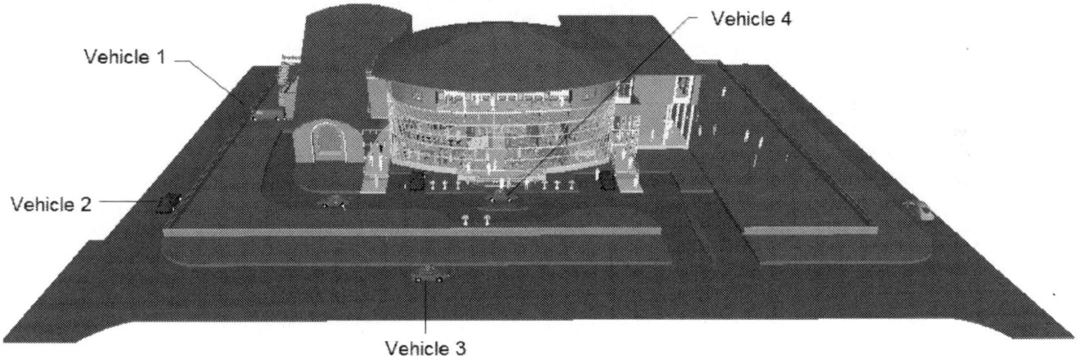

Figure 6-85 *The Building model*

Answers to Self-Evaluation Test

1. Edit Keyframe, **2.** On Key Press, **3.** Play Animation, **4.** Show Viewpoint, **5.** Enable Scripts, **6.** T, **7.** F, **8.** F, **9.** T, **10.** T

Chapter 7

Quantification

Learning Objectives

After completing this chapter, you will be able to:

- *Understand the concept of Quantification*
- *Use various options in the Quantification Workbook window*
- *Use Item Catalog*
- *Use Resource Catalog*
- *Perform model and 2D takeoff*

INTRODUCTION

In Navisworks, quantification is the process of analyzing drawings and preparing a list of all the materials that are required to complete the design. The list of materials includes the material quantities and material types.

In this chapter, you will learn to carry out the take off process by using the **Quantification Workbook** window.

QUANTIFICATION WORKBOOK WINDOW

The **Quantification Workbook** window contains various tools that are used for estimating the material quantities and their types, measuring areas, and counting building components. Instead of calculating the takeoff manually, you can use this feature for the estimation which helps in saving time as well as resources that are used for estimating the material quantities. To perform the take off, first you need to invoke the **Quantification Workbook** window.

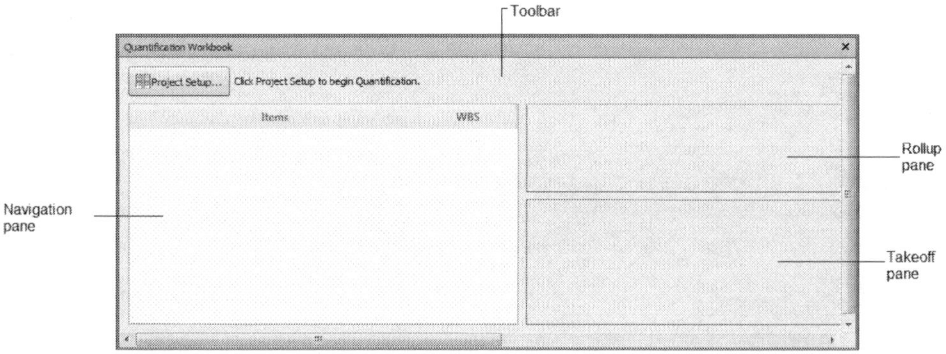

***Figure 7-1** Various components in the **Quantification Workbook** window*

Creating a Project

In Quantification, a project is a collection of files and takeoff items that are used for estimating material quantities. In Navisworks, there are three predefined project templates that you can use for takeoff calculations. These project templates are also known as catalogs.

Item Catalog

The **Item Catalog** window contains a list of takeoff items. You can also define resources in this window. To display the **Item Catalog** window, choose the **Catalog** button from the Toolbar in the **Quantification Workbook** window; a drop-down list will be displayed. Select the **Item Catalog** check box from the list; the **Item Catalog** window will be displayed, as shown in Figure 7-2. Figure 7-2 shows various components of the **Item Catalog** window.

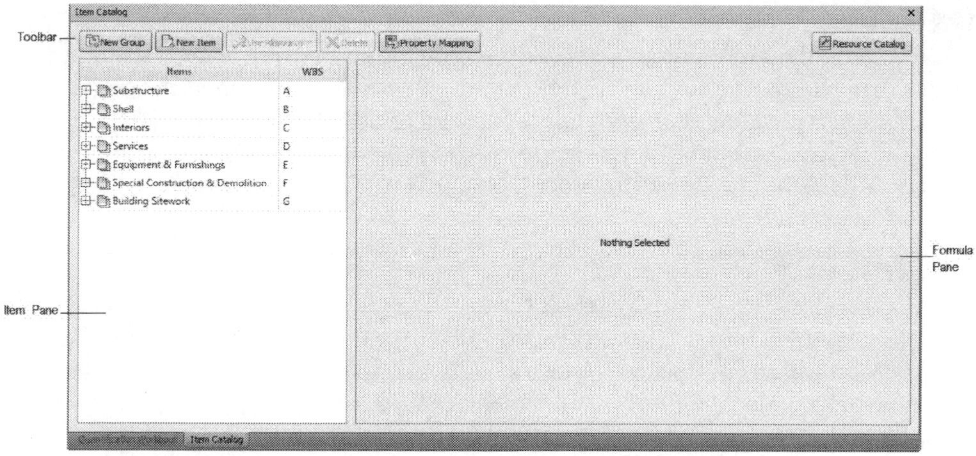

Figure 7-2 Various components of the ***Item Catalog*** *window*

Resource Catalog

The **Resource Catalog** contains a list of all the resources for a project. This resource list may include materials, equipments, or structural components. The various components of the resource window is shown in Figure 7-3.

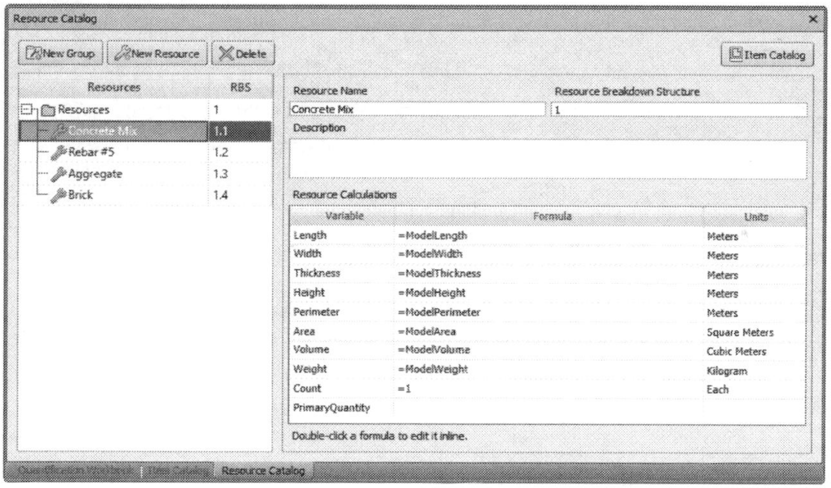

Figure 7-3 Various components of the ***Resource Catalog*** *window*

TAKEOFF METHODS

In Quantification, there are two takeoff methods: Model Takeoff and Virtual Takeoff. These two methods are also known as Automatic Takeoff and Manual Takeoff, respectively.

Model Takeoff

In the Model Takeoff method, the objects included in the files are estimated. Using this method, you can perform the takeoff for the objects that have properties mapped in Navisworks.
The **Take off to Selected Catalog Item** option is used to take off the selected objects to the item selected in the Navigation pane. This option will be available only when you choose an item in the Navigation pane of the **Quantification Workbook** window.

Virtual Takeoff

The Virtual Takeoff method is used to takeoff those objects that are not linked to a model object or the objects which contain no properties. You can use this method to takeoff the objects that have geometry but no properties.

2D Takeoff

The 2D Takeoff method is used to perform takeoffs such as calculation of area, length, or count of objects in a 2d sheet. This method helps saving time and effort required in manual calculation. To perform 2D takeoff, you need to load the required 2D plan. For this, you can use native or scanned DWF files.

TUTORIALS

General instructions for downloading tutorial files:

1. Download the *c07_navisworks_2024_tut* zip file for this tutorial from *https://www.cadsofttech.com*. The path of the file is as follows: ***Books > Civil/GIS Books > Autodesk Navisworks > Exploring Autodesk Navisworks 2024 for Novices (Learn By Doing) > Tutorial Files.***

2. Now, save and extract the downloaded folder at the following location:
C:\ nws_2024\c06_nws_2024_tut

Note
The default unit system used in the tutorials is metric. To change the unit to imperial, select the required unit from ***Options Editor > Interface > Display Units****.*

Tutorial 1 Model Takeoff

In this tutorial, you will open *c07_navisworks_2024_tut1* file and perform quantity takeoff of various building components by using the options available in the **Quantification Workbook** window. **(Expected time: 45 min)**

The following steps are required to complete this tutorial:

a. Open the *c07_navisworks_2024_tut1* file.
b. Display the **Quantification Workbook** window.
c. Set up the project.
d. Perform the takeoff.
e. Save the project.

Opening the File

In this section, you will open the file to be used in this tutorial.

1. Choose the **Open** button from the Quick Access Toolbar; the **Open** dialog box is displayed.

2. In this dialog box, browse to the following location:
   C:\nws_2024\c07_nws_2024_tut

3. Select **Navisworks File Set (*.nwf)** from the **Files of type** drop-down list.

4. Select the file **c07_navisworks_2024_tut1** from the displayed list of files; the file name appears in the **File name** edit box.

5. Next, choose the **Open** button in the dialog box; the model is displayed in the Scene View, as shown in Figure 7-4.

Figure 7-4 *Model displayed in the Scene View*

Displaying the Quantification Workbook Window

In this section, you will display the **Quantification Workbook** window.

1. Choose the **Quantification** tool from the **Tools** panel in the **Home** tab; the **Quantification Workbook** window is displayed, as shown in Figure 7-5.

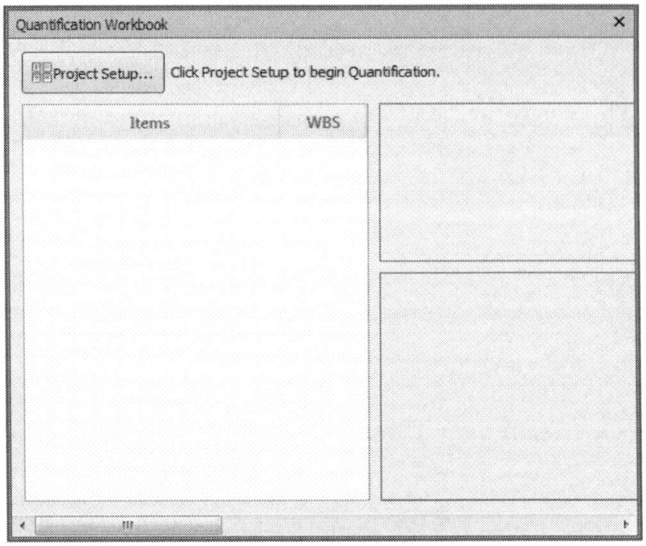

*Figure 7-5 The **Quantification Workbook** window*

Setting up the Project

In this section, you will browse to the catalog and set up a new Quantification session.

1. Choose the **Project Setup** button available at the top of the **Quantification Workbook** window, refer to Figure 7-5; the **Quantification Setup Wizard** dialog box is displayed, as shown in Figure 7-6.

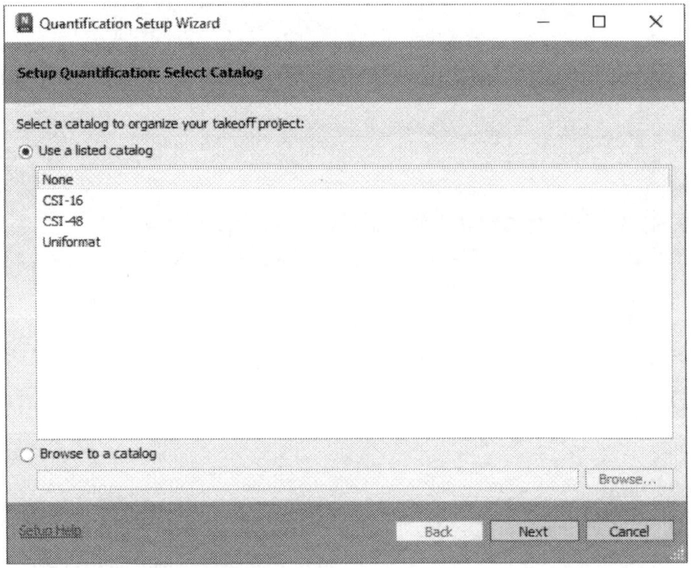

*Figure 7-6 The **Quantification Setup Wizard** dialog box*

Note
*If you are starting quantification for the first time, the **Quantification Getting Started Wizard** message box is displayed. In this message box, you can choose the **Never** button.*

2. In the **Setup Quantification: Select Catalog** page of the wizard, select the **Browse to a catalog** radio button; the **Browse** button is enabled.

3. Next, choose the **Browse** button; the **Open** dialog box is displayed.

4. In the **Open** dialog box, browse to the following location:
   C:\nws_2024\c07_nws_2024_tut

5. Select the **Quantification** xml file displayed in the file list.

6. Next, choose the **Open** button; the path of the selected catalog is displayed in the **Browse** edit box.

7. Now, choose the **Next** button in the **Setup Quantification: Select Catalog** of the wizard; the **Setup Quantification: Ready to Create Database** page is displayed, as shown in Figure 7-7.

8. Choose the **Finish** button; the catalog is added to the Navigation pane of the **Quantification Workbook** window, refer to Figure 7-7.

Note
For calculating the takeoff in Imperial units, use the Quantification_Imperial xml file from the C:\nws_2024\c07_nws_2024_tut location.

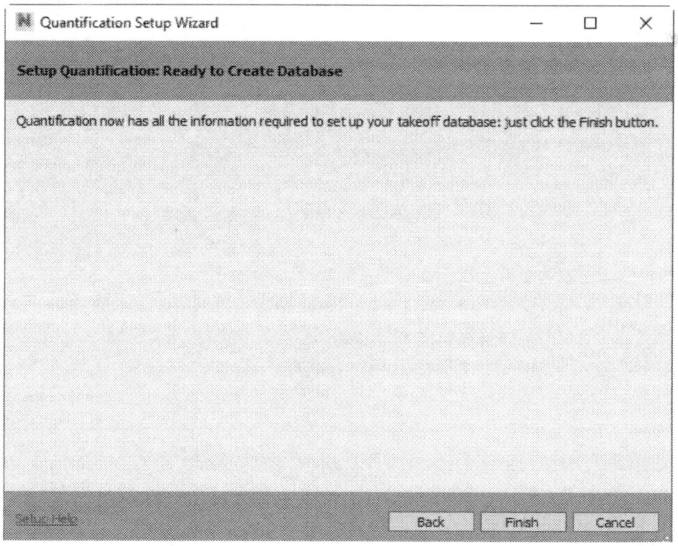

Figure 7-7 *The catalog added to the Navigation pane of the **Quantification Workbook** window*

Performing Quantity Takeoff

In this section, you will calculate the quantity takeoff of various structural components.

1. Expand the **Substructure** node under the Items column in the Navigation pane of the **Quantification Workbook**.

2. Expand **Substructure > Foundations > Standard Foundations > Rectangular Foundation(0)**, refer to Figure 7-8.

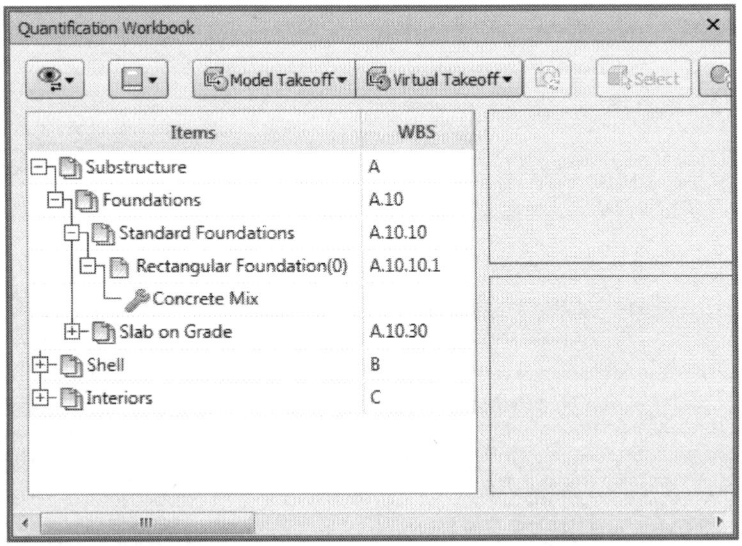

Figure 7-8 *The folders expanded in the Navigation pane of the window*

3. Now, choose the **Selection Tree** button from the **Select & Search** panel of the **Home** tab; the **Selection Tree** window is displayed.

4. Expand the **tut 2.nwc** folder in the **Selection Tree** window.

5. Next, expand the **Level 3 > Structural Foundations > Footing-Rectangular > 1800x1200x450** in the **Selection Tree** window, refer to Figure 7-9.

6. Now, select the desired foundations in the **Selection Tree** window, as shown in Figure 7-9.

7. Select the **Rectangular Foundation** item in the Navigation pane of the **Quantification Workbook**; its summary is displayed in the Rollup pane of the window.

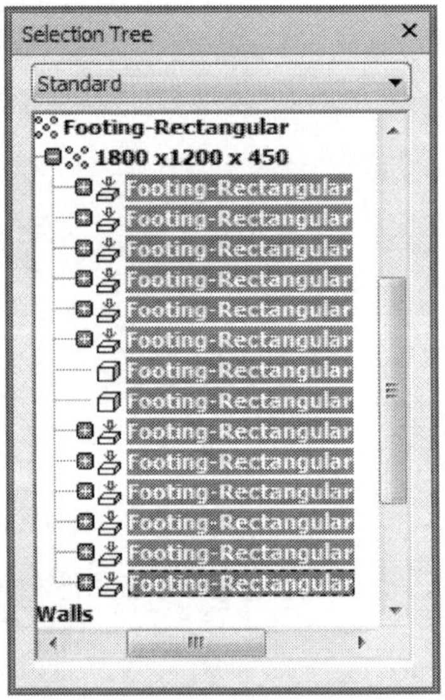

Figure 7-9 *Selecting foundations in the **Selection Tree** window*

8. Next, choose the **Model Takeoff** button in the Toolbar of the **Quantification Workbook**; a drop-down list is displayed.

9. Select the **Take off to: Rectangular Foundation** option from the list; the take off results are displayed in the Takeoff pane of the **Quantification Workbook**, as shown in Figure 7-10.

Note

The appearance of the selected objects changes after the takeoff.

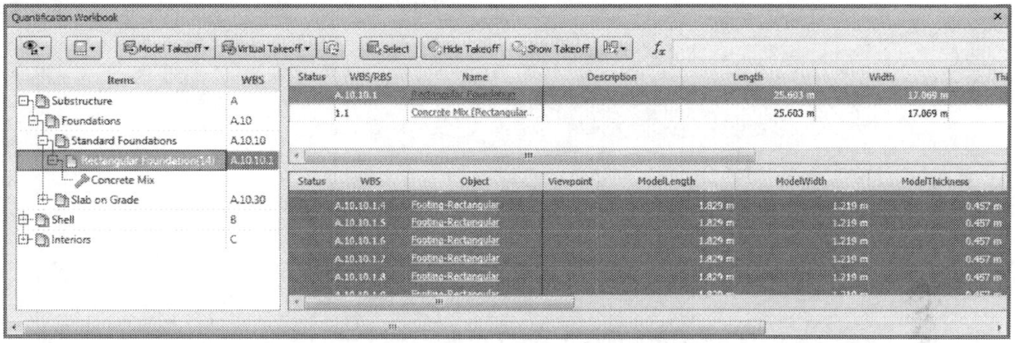

Figure 7-10 *Take off results displayed in the Takeoff pane of the window*

10. Rotate the model and then choose the **Select** tool from **Home > Select&Search > Select** drop-down and click on the foundation slab; the slab is highlighted in blue, as shown in Figure 7-11.

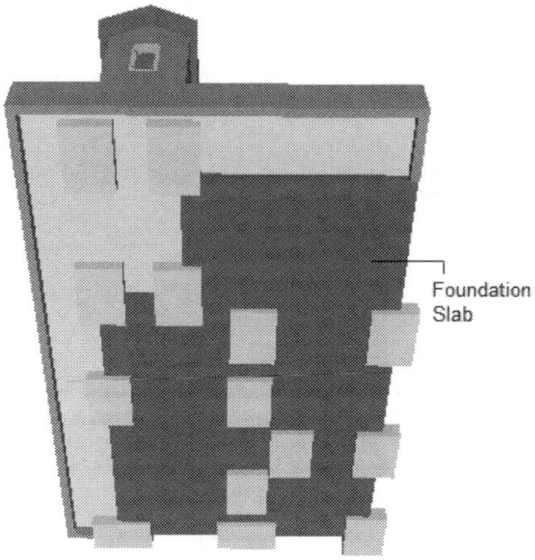

Figure 7-11 *Foundation slab selected in the Scene View*

11. Select the **Foundation Slab** item under the **Slab on Grade** node in the Navigation pane of the **Quantification Workbook** window, as shown in Figure 7-12.

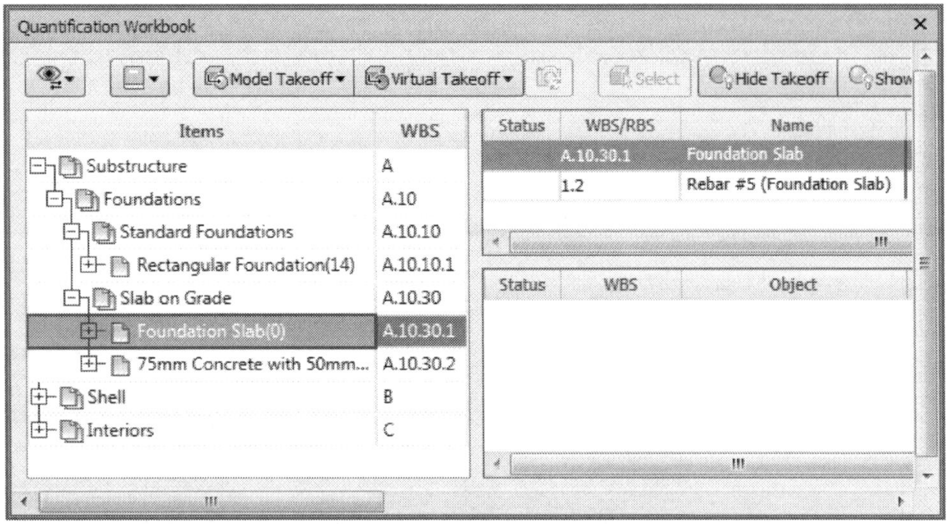

Figure 7-12 *Selecting the **Foundation Slab** item in the Navigation pane of the window*

12. Repeat the procedure followed in steps 8 and 9 to calculate the model takeoff of the selected object; the takeoff result is displayed in the Takeoff pane of the window.

13. Now, select the **Floor** slab in the Scene View; the slab is highlighted in blue, as shown in Figure 7-13.

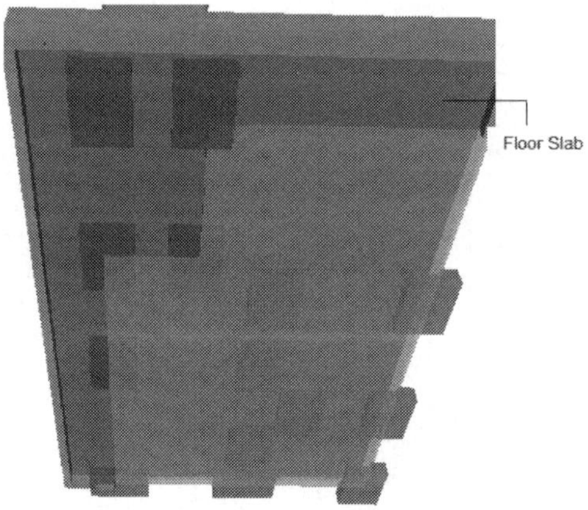

Figure 7-13 *The slab selected in the Scene View*

14. Select the **75mm Concrete with 50mm Metal Deck** item in the Navigation pane of the **Quantification Workbook** window, as shown in Figure 7-14.

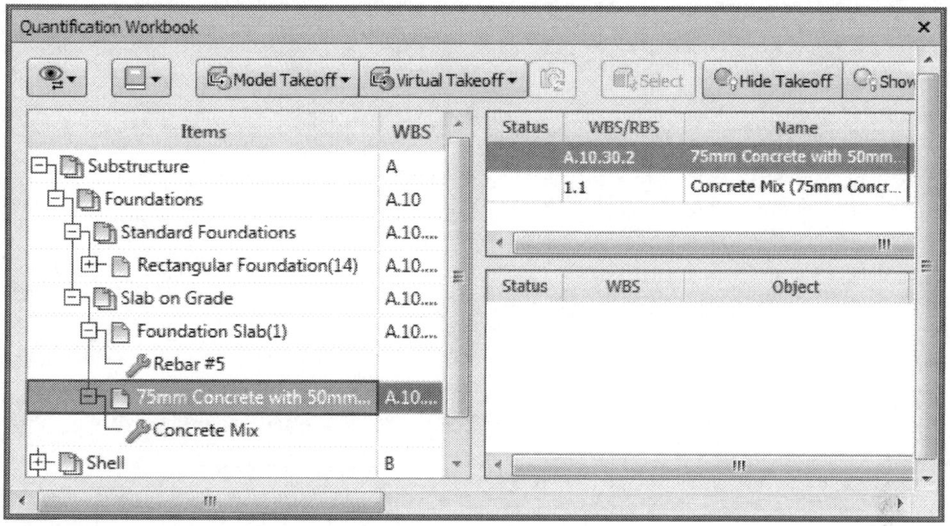

Figure 7-14 *Selecting* ***75mm Concrete with 50mm Metal Deck*** *in the window*

15. Repeat the procedure followed in steps 8 and 9 to calculate the model takeoff of the selected object; the takeoff result is displayed in the Takeoff pane of the window.

16. Select the structural columns available under the **Level 3** and **Level 5** nodes in the **Selection Tree** window, as shown in Figure 7-15.

Note

You can use the CTRL key to select multiple items in the ***Selection Tree*** *window.*

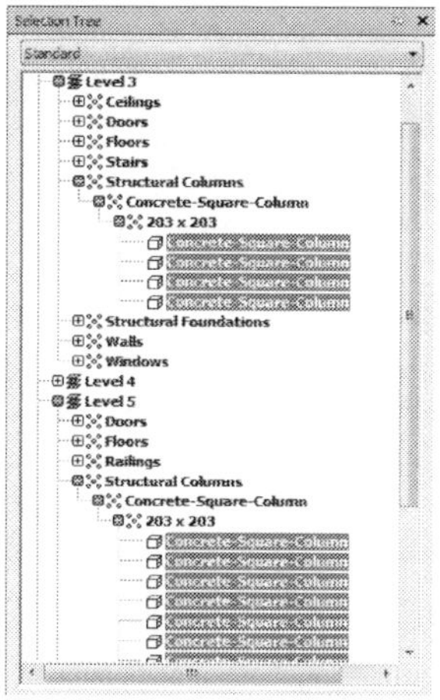

Figure 7-15 *Selecting columns in the* ***Selection Tree*** *window*

17. Next, select the **Columns** item from **Shell > Superstructure > Floor Construction** in the Navigation pane of the **Quantification Workbook** window, as shown in Figure 7-16.

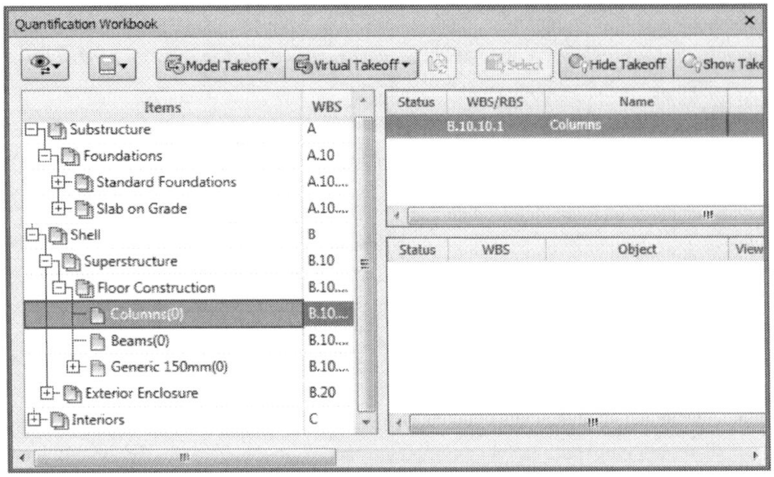

Figure 7-16 *Selecting columns in the Navigation pane of the window*

18. Repeat the procedure followed in steps 8 and 9 to calculate the model takeoff for the selected object; the takeoff result is displayed in the Takeoff pane of the window.

19. Next, in the **Selection Tree** window, select all the members by pressing the CTRL key under **Level 5 > Structural Framing > Concrete-Rectangular Beam > 203*304** node, as shown in Figure 7-17.

20. Select the **Beams** item under the **Shell** group in the Navigation pane of the **Quantification Workbook** window, as shown in Figure 7-18.

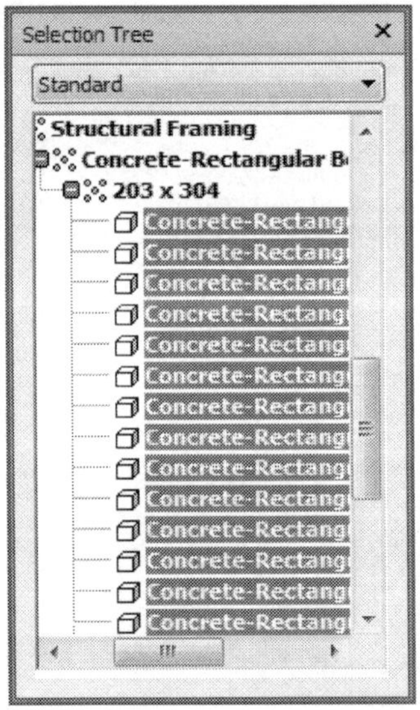

Figure 7-17 *Selecting beams in the* ***Selection Tree*** *window*

21. Repeat the procedure followed in steps 8 and 9 to calculate the model takeoff for the selected object.

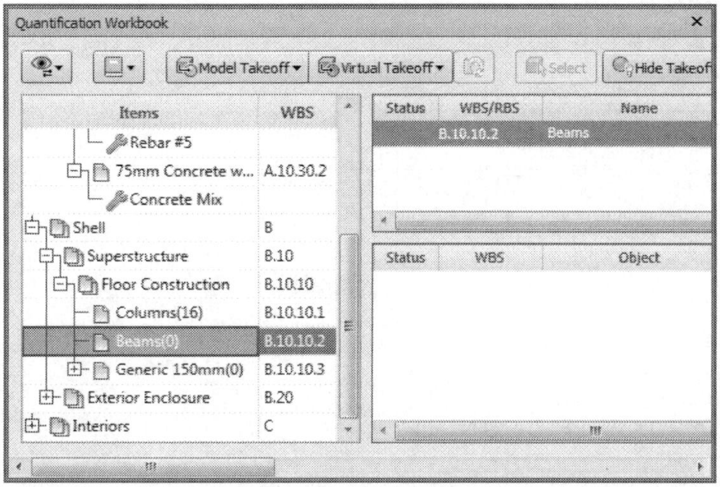

Figure 7-18 Selecting ***Beams*** *item in the Navigation pane of the window*

22. Select the floor of the model in the Scene View; the selected object is highlighted in blue color, as shown in Figure 7-19.

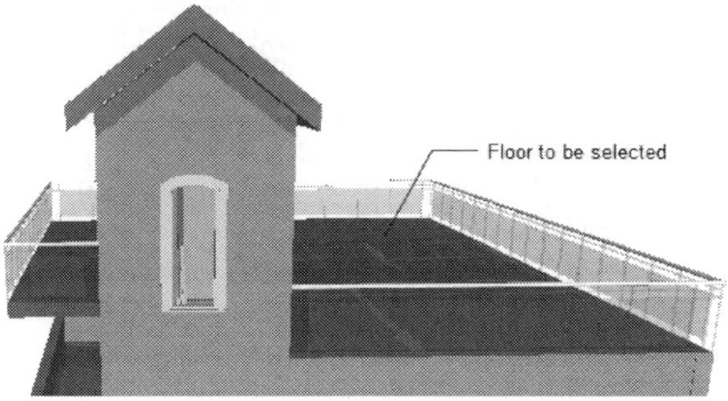

Figure 7-19 *Selecting the floor in the model*

23. Select the **Generic 150mm** item from **Shell > Superstructure > Floor Construction** in the Navigation pane of the workbook.

24. Repeat the procedure followed in steps 8 and 9 to calculate the model takeoff the selected object.

Exporting Quantities to Excel

In this section, you will export the takeoff quantities to an excel file.

1. In the Navigation pane of the **Quantification Workbook** window, select **Rectangular Foundations(14)**.

2. Press the CTRL key and select **Foundation Slab(1)**, **75mm Concrete with 50mm Metal Deck slab(1)**, **Columns(16)**, **Beams(21)**, and **Generic 150mm(1)**, as shown in Figure 7-20.

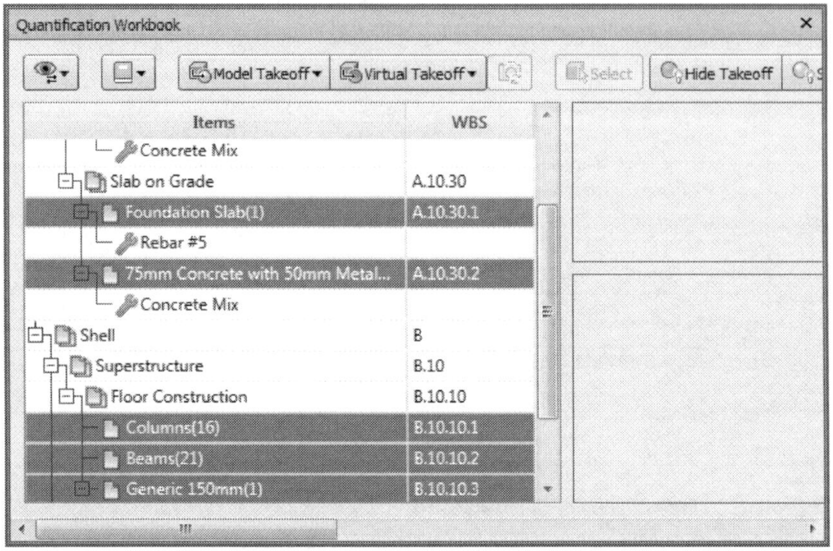

Figure 7-20 *Selecting items in the Navigation pane of the* ***Quantification Workbook*** *window*

3. Next, select the **Export Selected Quantities to Excel** option from the **Import/export Catalogs and export Quantities** drop-down list located at the extreme right in the **Quantification Workbook** window; the **Export Quantities to Excel** dialog box is displayed.

4. In this dialog box, browse to the following location: *C:\Navisworks_2024\c07_tutorial*.

5. Specify the name **Takeoff Quantities** in the **File name** edit box, as shown in Figure 7-21 and choose the **Save** button; the file is saved at the specified location and the **Export Quantities to Excel** message box is displayed.

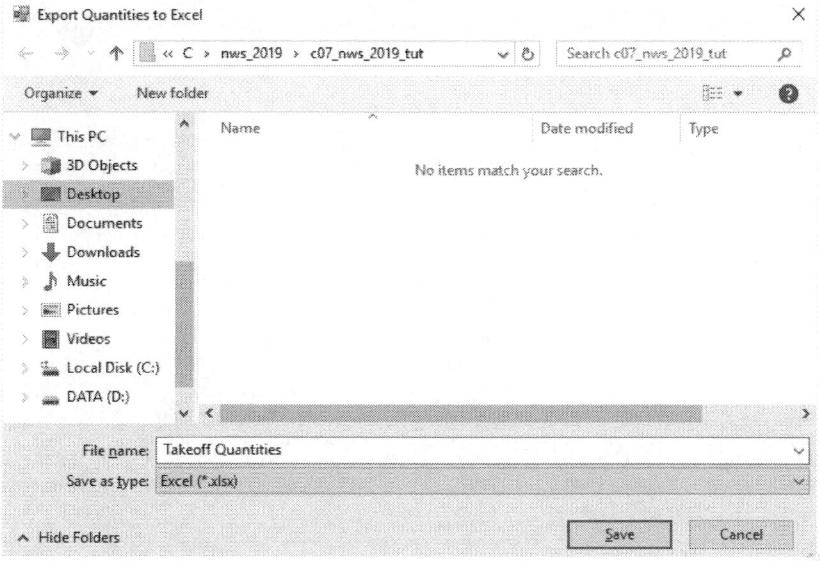

Figure 7-21 *The* ***Export Quantities to Excel*** *dialog box*

6. In this message box, choose the **Yes** button; the takeoff quantities are exported and opened in an excel file, as shown in Figure 7-22.

Saving the Project

In this section, you will save the project.

1. Before saving, choose the **Appearance** button in the Toolbar of the **Quantification Workbook** window; a drop-down list is displayed. 

2. Select the **Reapply Original Model Appearance** option from the list; the original appearance is applied to the model.

3. To save the project with the current view, choose **Save As** from the Application Menu; the **Save As** dialog box is displayed.

4. Browse to the *Navisworks_2024* folder and enter **c07_navisworks_2024_tut01** in the **File name** edit box. Now, select the **Navisworks File Set (*.nwf)** file format from the **Save as type** drop-down list, and choose the **Save** button.

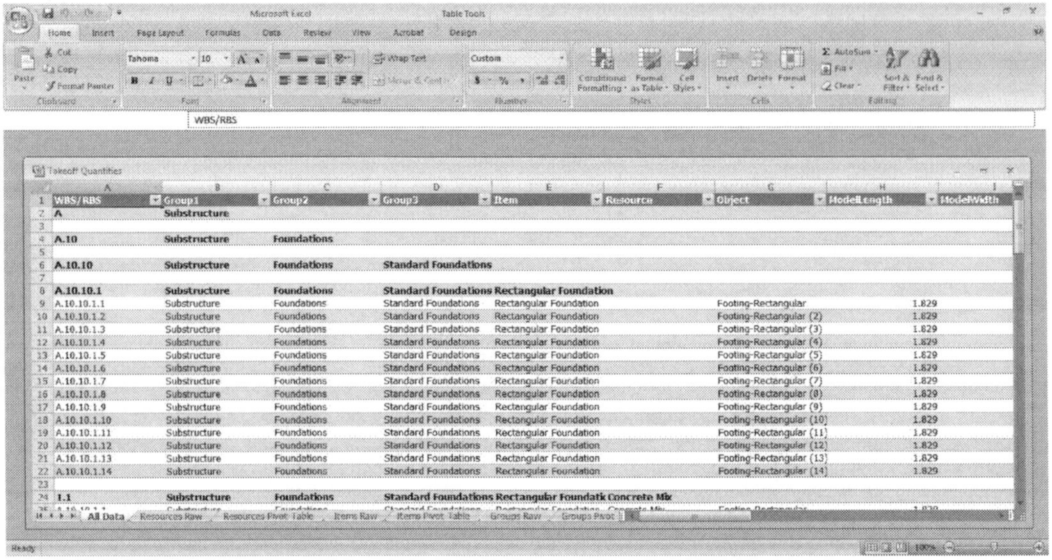

Figure 7-22 The takeoff quantities opened in an Excel file

Tutorial 2 2D Takeoff

In this tutorial, you will open the *c07_navisworks_2024_tut2* file and perform 2D takeoff on the floor plan by using the options available in the **Quantification Workbook** window.

(Expected time: 45 min)

The following steps are required to complete this tutorial:

a. Open the *c07_navisworks_2024_tut2* file.
b. Display the **Project Browser** window and import the 2D sheet.
c. Set up the project in the **Quantification Workbook** window.
d. Perform the takeoff.
e. Save the project.

Opening the File

In this section, you will open the file to be used in this tutorial.

1. Choose the **Open** button from the Quick Access Toolbar; the **Open** dialog box is displayed.

2. In this dialog box, browse to the following location:
   C:\Navisworks_2024\c07_tutorial

3. Select **Navisworks File Set (*.nwf)** from the **Files of type** drop-down list.

4. Select the file **c07_navisworks_2024_tut2** from the displayed list of files; the file name appears in the **File name** edit box.

5. Next, choose the **Open** button in the dialog box; the model is displayed in the Scene View, as shown in Figure 7-23.

Figure 7-23 *Model displayed in the Scene View*

Importing the 2D Sheet

In this section, you will import the floor plan of the model using the **Sheet Browser** window.

1. Choose the **Sheet Browser** button available in the Status Bar; the **Sheet Browser** window is displayed.

2. In this window, choose the **Import Sheets & Models** button; the **Insert From File** dialog box is displayed. In this dialog box, select the DWF option.

3. In this dialog box, browse to the following location:
C:\Navisworks_2024\c07_tutorial

4. Select the *c07_tut2_floor plan.dwf* file from the displayed list of files. Choose the **Open** button; the file gets added and is displayed with the name **Model** in the **Sheet Browser** window.

5. Choose the **Next Sheet** arrow in the Status Bar; the 2D sheet is displayed in the Scene View, refer to Figure 7-24.

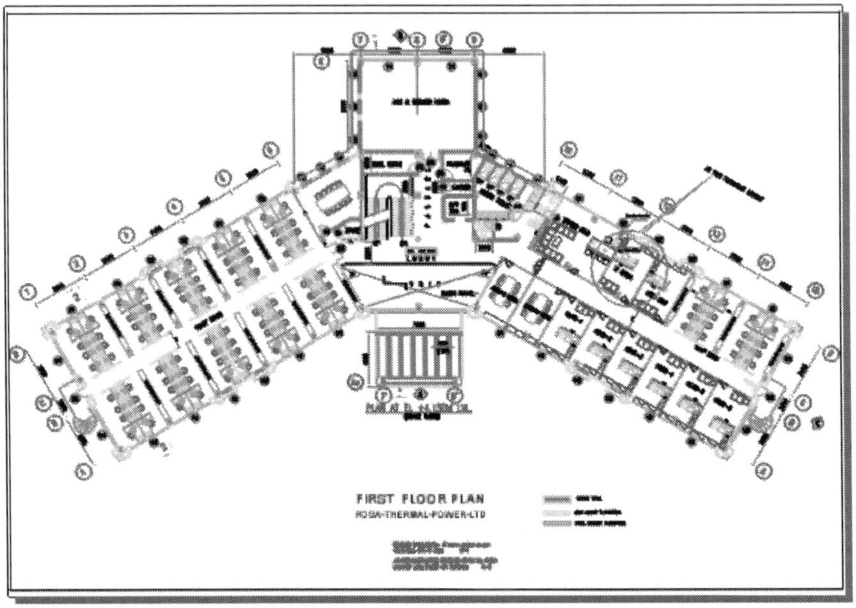

Figure 7-24 *2D Sheet displayed in the Scene View*

Setting Up the Project in the Quantification Workbook Window

In this section, you will invoke the **Quantification Workbook** window and import a quantification catalog to perform the takeoff.

1. Choose the **Quantification** tool from the **Quantification** panel of the **Home** tab; the **Quantification Workbook** window is displayed.

2. Next, choose the **Project Setup** button; the **Quantification Setup Wizard** is displayed. Select the **Browse to a catalog** radio button and choose the **Browse** button; the **Open** dialog box is displayed.

3. In the **Open** dialog box, browse to the following location:
   C:\Navisworks_2024\c07_tutorial

4. Select the *2D Takeoff.xml* file from the file lists and choose the **Open** button; the **Open** dialog box closes and then choose the **Next** button in the **Quantification Setup Wizard** dialog box; the **Setup Quantification: Ready to Create Database** page is displayed.

5. Choose the **Finish** button; the catalog is imported to the **Quantification Workbook** window, refer to Figure 7-25.

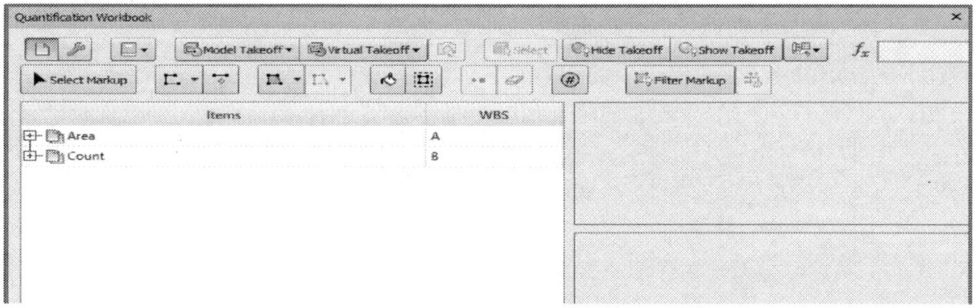

Figure 7-25 The 2D takeoff imported to the ***Quantification Workbook*** *window*

Performing the Takeoff

In this section, you perform the area and count takeoff on the 2D sheet.

1. In the Navigation pane of the **Quantification Workbook** window, expand the **Area** group and select the **Area of Chiller Room** item, refer to Figure 7-26.

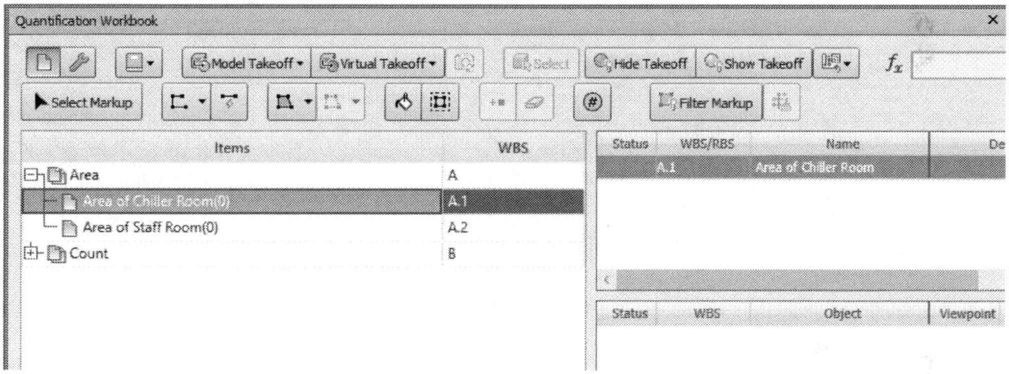

***Figure 7-26** The **Area of Chiller Room** item selected in the window*

2. Click in the Scene View and zoom in using the middle mouse button to view the chiller room, refer to Figure 7-27.

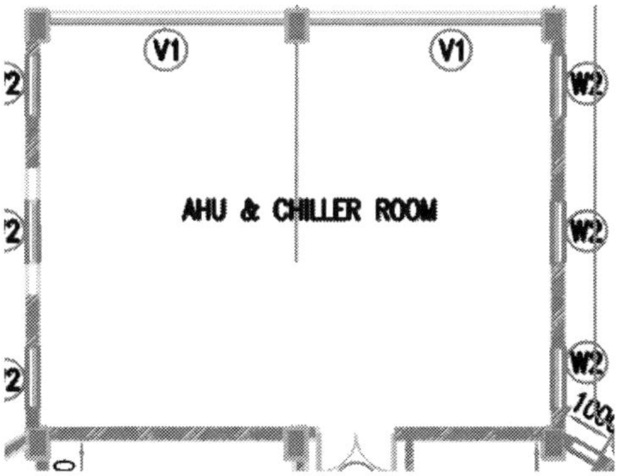

***Figure 7-27** The chiller room*

3. Select the Rectangle Area(A) option from the Area drop-down list in the Quantification Workbook window.

4. Place the cursor at the top left corner of the room and click when a green symbol appears, as shown in Figure 7-28. Notice that a white dot is displayed at the clicked point and a rectangle is attached with the cursor, refer to Figure 7-29.

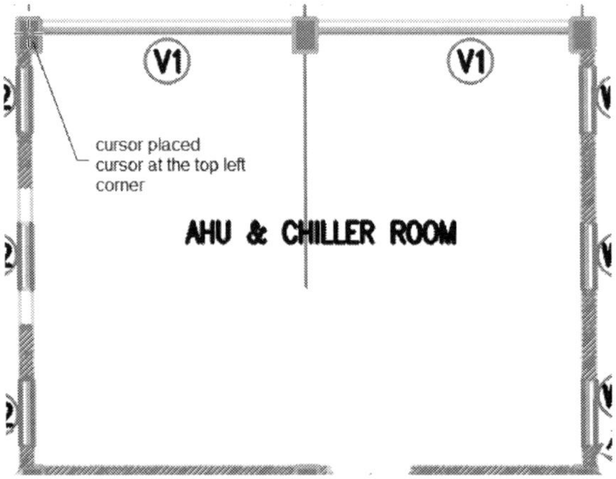

Figure 7-28 *Cursor placed at the corner*

5. Move the cursor downward and click at the bottom right corner of the room, refer to Figure 7-30.

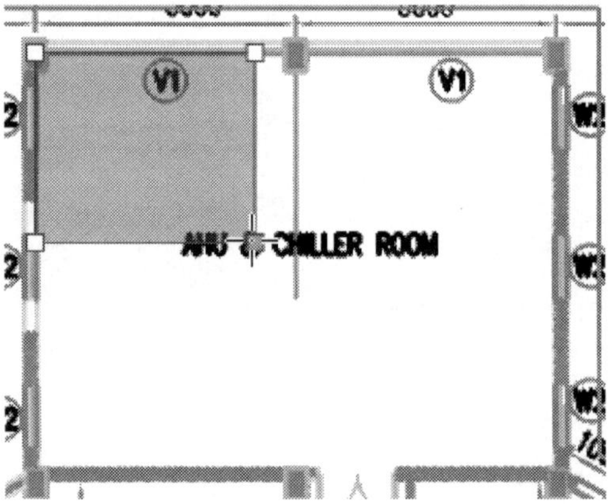

Figure 7-29 *Drawing area markup*

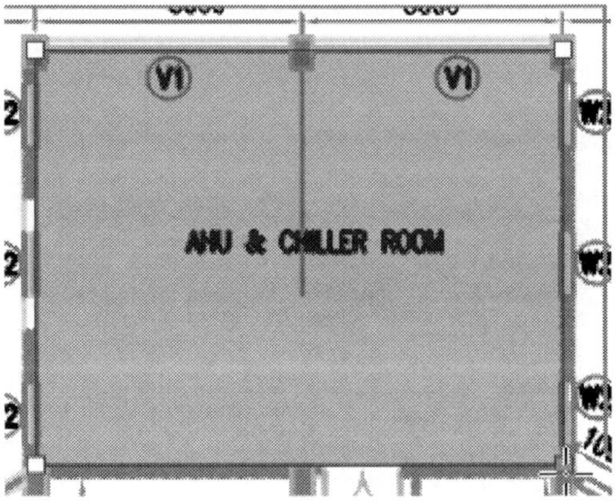

Figure 7-30 *Drawing area markup*

6. The area takeoff of the chiller room is displayed in the Takeoff pane of the **Quantification Workbook** window.

Now, you will perform the area takeoff for the staff room located in the left side of the plan.

7. Invoke the **Zoom** tool from the Navigation Bar and zoom in to view the staff room area, refer to Figure 7-31.

Figure 7-31 *Staff room in the plan*

8. Select the **Area of Staff Room** item in the Navigation pane of the **Quantification Workbook** window and select the **Area (A)** option from the **Area** drop-down list.

9. Place the cursor at the first left corner, refer to Figure 7-32. Click at that point and then move the cursor downward; a blue line is attached to the cursor, refer to Figure 7-32.

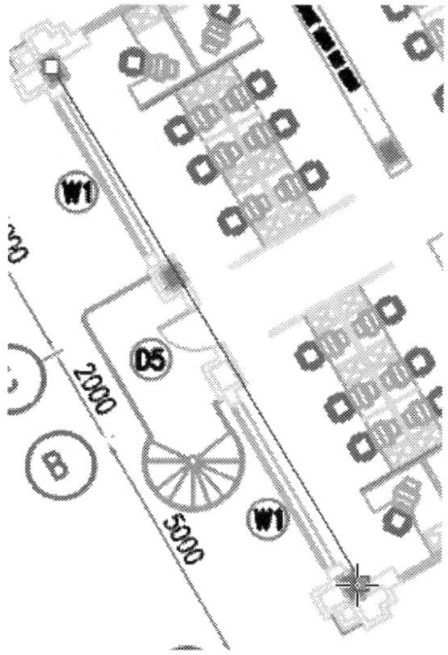

Figure 7-32 *Blue line attached to the cursor*

10. Click at the second corner and move the cursor in the right direction and click at the third corner, refer to Figure 7-33.

11. Move the cursor upward and click at the fourth corner. Next, right-click to complete the area markup, refer to Figure 7-34. The area takeoff is displayed in the Takeoff pane of the **Quantification Workbook** window.

Figure 7-33 *Drawing markup*

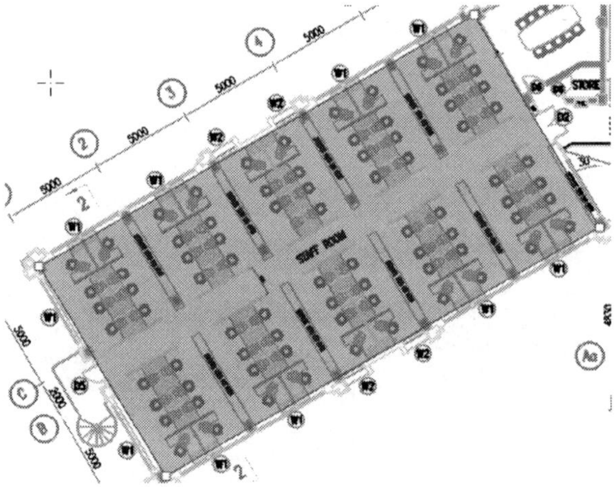

Figure 7-34 *Complete markup*

12. Choose the **Backout** button; a drop-down list is displayed. Now, select the **Backout (B)** option from the list.

13. Place the cursor at the point marked as A, refer to Figure 7-35. Click when a green symbol appears.

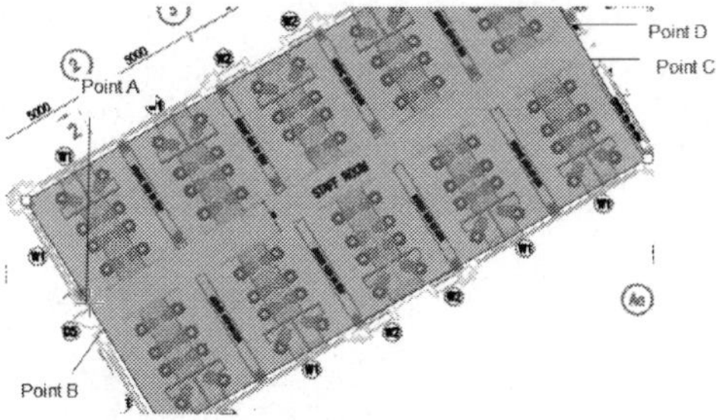

Figure 7-35 *Cursor placed at point A*

14. Move the cursor downward and click at point B. Similarly, click at points C and D and right-click to complete; the backout markup is created, refer to Figure 7-36 and the resulting area is displayed in the Takeoff pane of the **Quantification Workbook** window.

Note
You can deselect a markup by clicking in the Scene View.

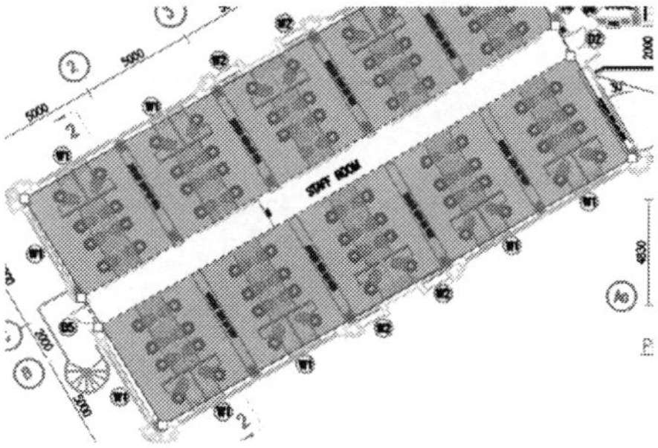

Figure 7-36 *Backout Created*

15. Choose the **Catalog** button from the Quantification toolbar; a drop-down list is displayed. Select the **Item Catalog** option from the list; the **Item Catalog** window is displayed. 

16. In this window, expand the **Count** group and select the **Chairs** item; the properties are displayed in the Formula pane.

17. In the **Object Appearance** area of the Formula pane, choose the **Color** button; the **Color** palette is displayed. Select the pink color and choose the **OK** button and then close the **Item Catalog** window. 

18. Expand the **Count** group and select the **Chair** item in the Navigation pane of the **Quantification Workbook** window.

19. Choose the **Count** button from the toolbar in the **Quantification Workbook** window. 

20. Click at the chairs in the staff room one by one to count the objects; the count takeoff is displayed in the Takeoff pane of the **Quantification Workbook** window and the number of chairs are displayed in the Navigation pane. Figure 7-37 shows the chairs marked with pink dots as count takeoff.

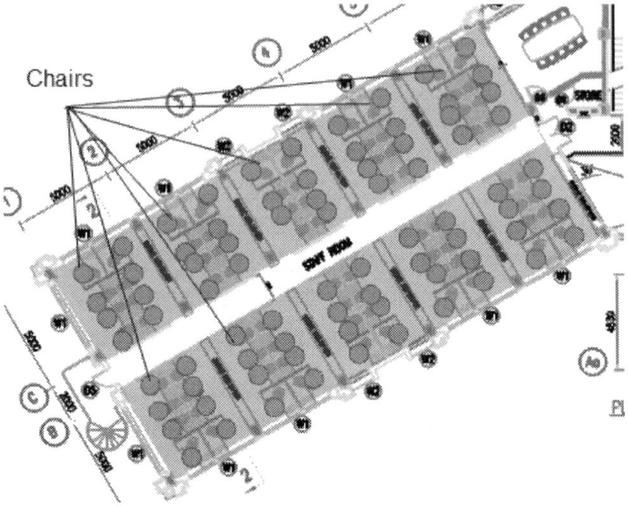

***Figure 7-37** Chairs marked as count takeoff*

Saving the Project

In this section, you will save the project.

1. To save the project with the current view, choose **Save As** from the Application Menu; the **Save As** dialog box is displayed.

2. Browse to the *nws_2024\ c07_nws_2024_tut* folder and enter **c07_navisworks_2024_tut02** in the **File name** edit box. Now, select the **Navisworks File Set (*.nwf)** file format from the **Save as type** drop-down list, and choose the **Save** button.

SELF-EVALUATION TEST

Answer the following questions and then compare them to those given at the end of this chapter:

1. The **Quantification Workbook** window can be invoked from the __________ panel of the __________ tab.

2. The __________ option is used to apply the original appearance of the model to the objects after performing the takeoff.

3. The __________ button is used to select the markups in the sheet.

4. The catalog items can be imported/exported from the __________ and __________ files.

5. The **Polyline (L)** option is used to draw markups for the area which are non-standard in shape. (T/F)

6. Viewpoints cannot be added to the takeoff objects. (T/F)

7. Model Takeoff is a method of quantifying model objects. (T/F)

8. Units of the takeoff results cannot be changed. (T/F)

9. The **Item Catalog** window contains a list of all the resources for the project. (T/F)

10. The **Master Resource List** dialog box contains a list of existing resources in a project.(T/F)

REVIEW QUESTIONS

Answer the following questions:

1. Which of the following options is used to filter the markup of the takeoff items?

   a) **Select Markup** b) **Filter Markup**
   c) **Count** d) None of these

2. Which of the following options is used for excluding an area from the area markup?

   a) **Polyline (L)** b) **Backout**
   c) **Quick Line** d) **Model Takeoff**

3. Which of the following options is used for creating a linear takeoff?

   a) **Quick Box** b) **Quick Line**
   c) **Erase** d) None of these

4. Which of the following options is used to create a blank takeoff?

   a) **Model Takeoff** b) **Virtual Takeoff**
   c) **Update** d) **Change Analysis**

5. Which of the following options is used to display all the items in the Quantification Workbook window?

   a) **Item View** b) **Resource View**
   c) **2D Takeoff** d) None of these

6. Navisworks does not allow you to takeoff objects from the 2D sheet. (T/F)

7. You cannot change the formula for any of the takeoff methods. (T/F)

8. The **Count** button is used to create the count takeoff of the objects. (T/F)

9. You cannot remove any unwanted geometry from the markup. (T/F)

10. The color of the takeoff objects cannot be changed. (T/F)

EXERCISES

Exercise 1 Model Takeoff

Download and open the *c07_navisworks_2024_ex1* file from *https://www.cadsofttech.com*. Calculate the Quantity takeoff of the Rectangular-Footings, Wall Foundation, Columns, Beams, and Roof Slab. Figure 7-38 shows the model to be used for this exercise. **(Expected time: 45 min)**

The following steps are required to complete this exercise:

a) Open the *c07_navisworks_2024_ex1* file.
b) Use the catalog Quantification_Exer for metric and Quantification_ Ex_imp for imperial unit.
c) Perform the takeoff of beams, columns, foundation, footing, and roof slab.
d) Export the takeoff quantities to an Excel file.
e) Save the project as *c07_navisworks_2024_ex02*.

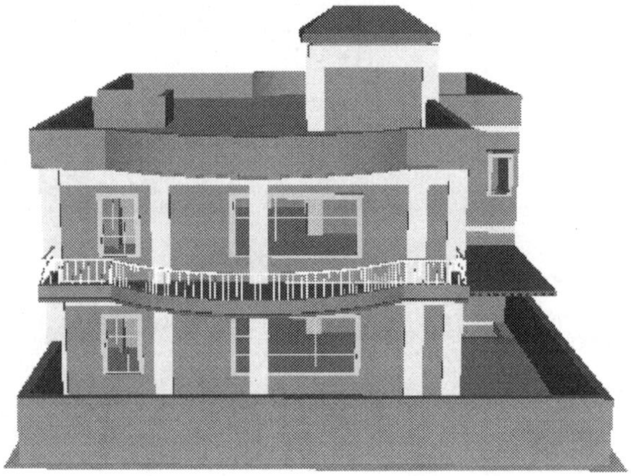

Figure 7-38 *The Residence Building*

Exercise 2 2D Takeoff

Download and open the *c07_navisworks_2024_ex2* file from *https://www.cadsofttech.com*. Calculate the Quantity takeoff of the Common Room and Bed Room. Figures 7-39 and 7-40 show the model and its floor plan to be used for this exercise. **(Expected time: 30 min)**

The following steps are required to complete this exercise:

a) Open the *c07_navisworks_2024_ex2* file.
b) Import the *c07_residence.dwf* file using the **Project Browser** window.
c) Open the loaded dwf file using the **Next Sheet** button from the Status Bar.
d) Invoke the **Quantification Workbook** window and import the 2D Takeoff_ex2 xml file.
e) Perform the area takeoff for the common room and bed room using the area takeoff tool.

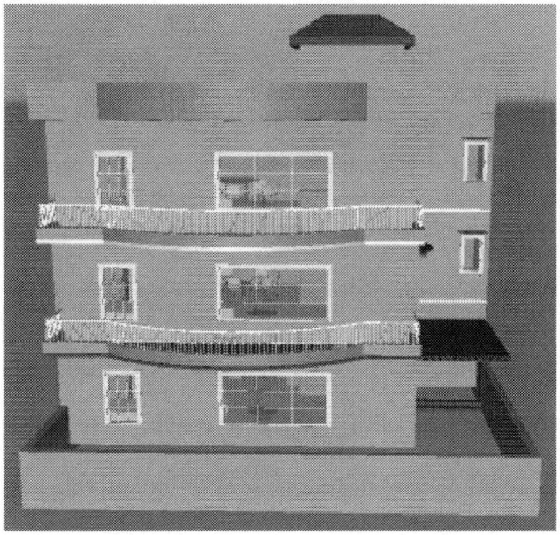

Figure 7-39 *The Residence building*

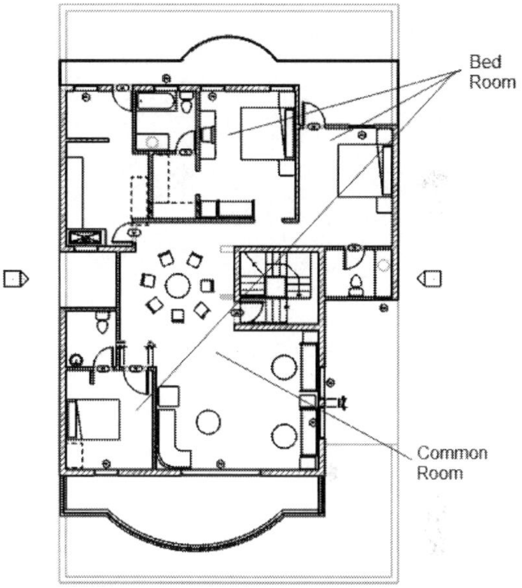

Figure 7-40 *Floor plan of the residence building*

Answers to Self-Evaluation Test

1. **Tools**, **Home**, **2**. **Reapply Original Model Appearance**, **3**. **Select Markup**, **4**. **Excel**, **XML**, **5**. T, **6**. F, **7**. T, **8**. F, **9**. F, **10**. T

Chapter 8

Clash Detection

Learning Objectives

After completing this chapter, you will be able to:

- *Understand the concept of clash detection*
- *Create a test*
- *Select objects for clash test*
- *Perform the clash test*
- *View clash results*
- *Generate a clash report*

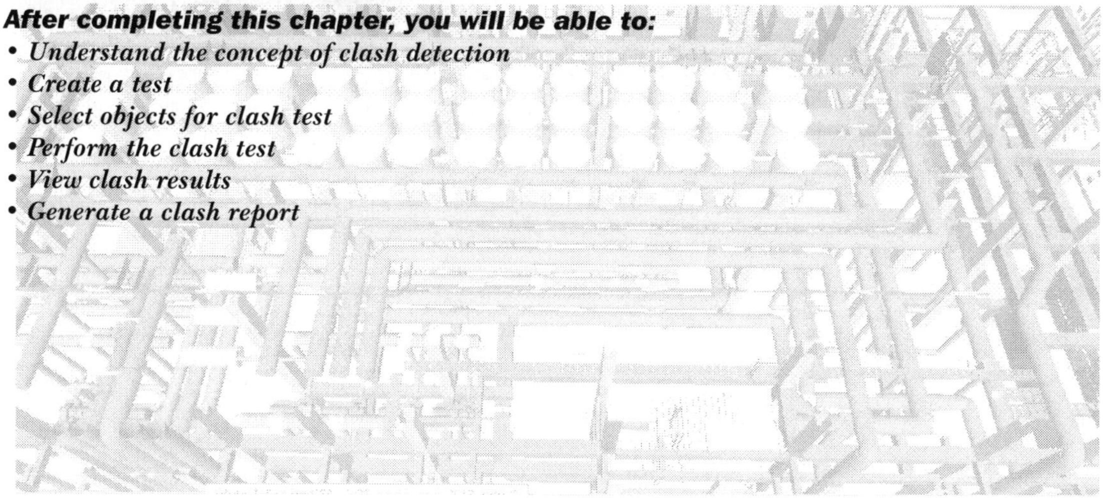

INTRODUCTION

Clash detection and coordination, though it is not an entirely new concept, has gained an impetus in the recent years. It has invariably played an important role in lowering down the construction costs. This concept had helped various project managers to control the huge cost overruns and limits the overall budget of construction. Clash detection has also helped to avert the unnecessary delays in the project due to lack of coordination between the stakeholders.

Before using BIM, the clashes were determined with the help of charts and drawings that were overlaid on a light table. This process was time consuming and was not accurate.

Moreover, clash detection and coordination help one to find the incompatibility and inconsistencies much before starting the actual construction. Hence, the clash detection process plays a crucial role in avoiding delays and saving time and resources.

CLASH DETECTIVE WINDOW

The **Clash Detective** window is a dockable window in which you can specify rules and options for performing clash tests. In this window, you can view and arrange results and can also generate clash reports.

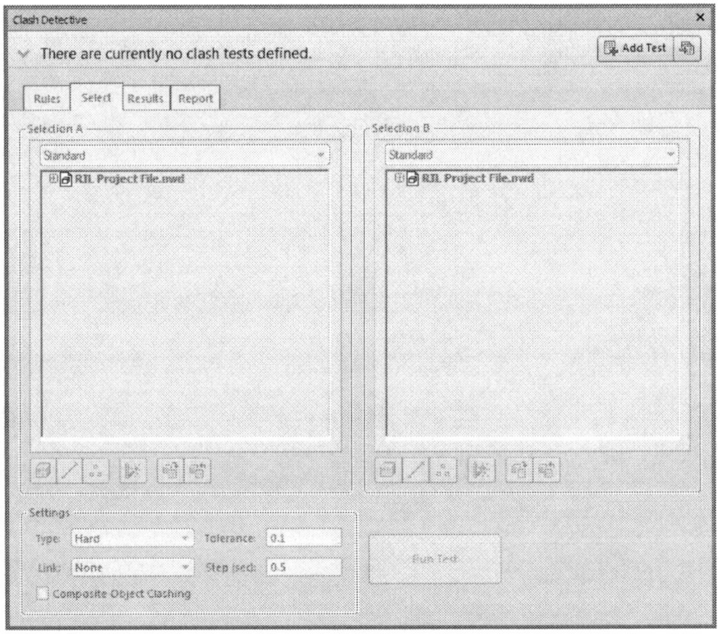

***Figure 8-1** The **Clash Detective** window*

Test Panel

The Test panel comprises of all the clash tests listed in a tabular format. In the **Clash Detective** window, the Test panel is not displayed by default. To display it, choose the **Add Test** button located at the top right corner in the window; the Test panel will be displayed with a name **Test1** under the **Name** column, as shown in Figure 8-2.

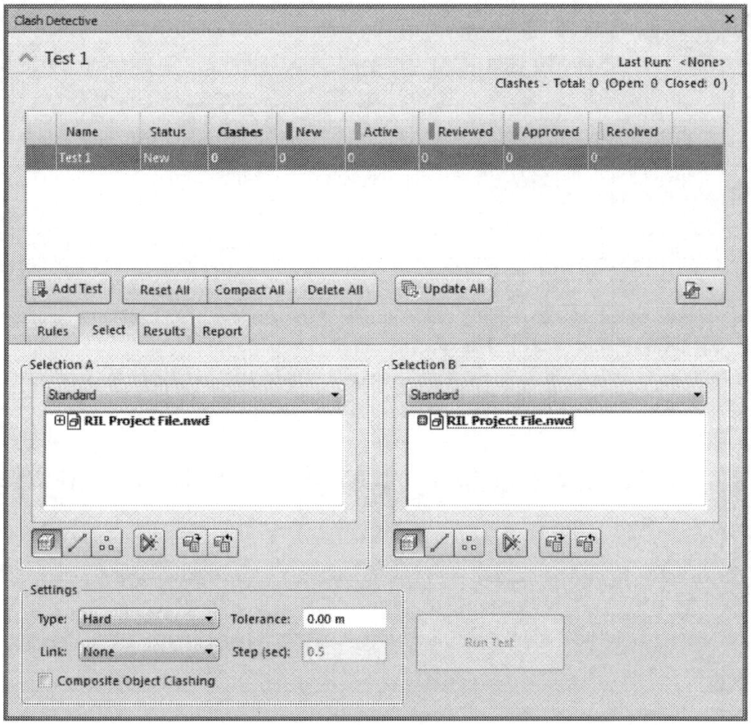

***Figure 8-2** The **Clash Detective** window with the Test panel displayed*

Rules Tab

When you choose the **Rules** tab in the **Clash Detective** window, a list of rules will be displayed, as shown in Figure 8-6. You can use these rules to define the situations in which Navisworks ignores clash between two objects.

Select Tab

The **Select** tab is chosen by default in the **Clash Detective** window. In this tab, you can define the clash test by selecting multiple sets of items at a time. Instead of testing the whole model against itself, a particular set of objects in the model will be tested. The **Select** tab comprise of the **Selection A** and **Selection B** areas at the top and the **Settings** area at the bottom.

Results Tab

In the **Results** tab, you will be able to view the clash results. This tab is divided into three areas: **Results View**, **Display Settings** panel, and **Items** panel.

Report Tab

In the **Report** tab, there are several options that are used for generating a report. The generated report will contain the details of all the clash results of the selected clash test. The **Report** tab comprises of three areas: **Contents**, **Include Clashes**, and **Output Settings**.

MANAGING CLASH TESTS

In this section, you will learn about merging and managing clash tests. You will learn about merging clash tests from different files, importing and exporting clash tests, and creating custom clash tests.

Merging Clash Tests

In Navisworks, when you merge different files into a single Navisworks file, all the clash data from different files will be combined together.

Creating Custom Clash Tests

If you are using a common set of clash tests in the multiple projects, then a custom clash test can be created which will be used everywhere.

LINKING CLASH TEST WITH TIMELINER AND OBJECT ANIMATION

In Navisworks, you can link the clash test with timeliner simulation and object animation. When you link the clash test with the timeliner, you are able to identify the clashes for the entire project. Similarly, by linking the clash test with the object animation, you can identify the clashes between moving objects.

Time-Based Clash Test

By linking clash detection with the timeliner simulation, you can identify the clashes at each step in the timeliner sequence. The scheduled task types such as temporary, demolished, or construct can be linked to the clash detection of cranes, bulldozer, and so on. This feature will help you in monitoring a project on a timely basis.

Soft Clashing

In Navisworks, you can also link an object animation scene created in the **Animator** window to a clash test to identify the clashes between moving objects.

Time-Based Soft Clashing

Before linking the clash test with the animated timeliner schedule, ensure that at least one animation scene has been linked to one of the tasks in the **TimeLiner** window. Next, in the **Select** tab of the **Clash Detective** window, define the clash test and select the **TimeLiner** option from the **Link** drop-down list. Specify the interval size in the **Step (sec)** edit box, and choose the **Run Test** button; the clash detective will check the clashes at each interval and the results will be displayed in the window. Now, view the clash results in the same way as for the Time-Based Clash Test.

CUSTOMIZING CLASH DETECTIVE WINDOW

You can also control the display of columns in the **Result** tab. To do so, right-click on the column header in the Results View area; a shortcut menu will be displayed. Choose the **Choose Columns** from the menu; the **Choose Columns** dialog box will be displayed. In the dialog box, select the check box corresponding to the column name that you want to display in the result table. Next, choose the **OK** button; the selected column will be displayed in the column table. Note that clearing the check box will remove the column from the Results View area of the **Results** tab.

CREATING CLASH REPORT

In Navisworks, several types of reports can be produced. You can share these reports with others. To create a clash report after running the clash test, choose the **Report** tab. In the **Contents** area of the **Report** tab, select the check boxes to be included in the report. In the **Include Clashes** area, select the required option from the **For Clash Groups, include** drop-down list for displaying the clash groups in the report. For example, if you select the **Group Headers Only** option then the clash report will include a summary of the clash groups and individual clashes which are not grouped. Select the status check boxes in the **Include these statuses** area for including the status of clash results. Select the type of report from the **Report Type** drop-down list. Select the report format from the **Report Format** drop-down list. Choose the **Write Report** button to write the report. When you choose this button, the **Save As** dialog box will be displayed. Browse to the folder and save the file in the desired report format such as .xml, .html. If you select the HTML report format, then the report will be created, as shown in Figure 8-3.

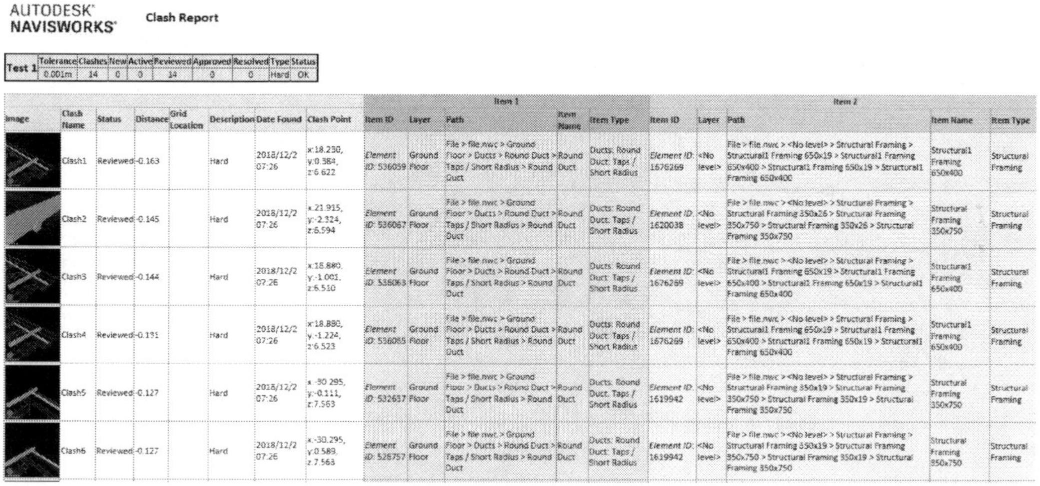
AUTODESK NAVISWORKS Clash Report

	Tolerance	Clashes	New	Active	Reviewed	Approved	Resolved	Type	Status
Test 1	0.001m	14	0	0	14	0	0	Hard	OK

								Item 1					Item 2				
Image	Clash Name	Status	Distance	Grid Location	Description	Date Found	Clash Point	Item ID	Layer	Path	Item Name	Item Type	Item ID	Layer	Path	Item Name	Item Type
	Clash1	Reviewed	-0.163		Hard	2018/12/2 07:26	x:18.230, y:0.384, z:6.622	Element ID: 536059	Ground Floor	File > file.nwc > Ground Floor > Ducts > Round Duct > Taps / Short Radius > Round Duct	Round Duct	Ducts: Round Duct: Taps / Short Radius	Element ID: 1676269	<No level>	File > file.nwc > <No level> > Structural Framing > Structural1 Framing 650x19 > Structural1 Framing 650x400 > Structural1 Framing 650x19 > Structural1 Framing 650x400	Structural1 Framing 650x400	Structural Framing
	Clash2	Reviewed	-0.145		Hard	2018/12/2 07:26	x:21.915, y:-2.324, z:6.594	Element ID: 536067	Ground Floor	File > file.nwc > Ground Floor > Ducts > Round Duct > Taps / Short Radius > Round Duct	Round Duct	Ducts: Round Duct: Taps / Short Radius	Element ID: 1620038	<No level>	File > file.nwc > <No level> > Structural Framing > Structural Framing 350x26 > Structural Framing 350x750 > Structural Framing 350x26 > Structural Framing 350x750	Structural Framing 350x750	Structural Framing
	Clash3	Reviewed	-0.144		Hard	2018/12/2 07:26	x:18.880, y:-1.001, z:6.510	Element ID: 536063	Ground Floor	File > file.nwc > Ground Floor > Ducts > Round Duct > Taps / Short Radius > Round Duct	Round Duct	Ducts: Round Duct: Taps / Short Radius	Element ID: 1676269	<No level>	File > file.nwc > <No level> > Structural Framing > Structural1 Framing 650x19 > Structural1 Framing 650x400 > Structural1 Framing 650x19 > Structural1 Framing 650x400	Structural1 Framing 650x400	Structural Framing
	Clash4	Reviewed	-0.131		Hard	2018/12/2 07:26	x:18.890, y:-1.224, z:6.523	Element ID: 536065	Ground Floor	File > file.nwc > Ground Floor > Ducts > Round Duct > Taps / Short Radius > Round Duct	Round Duct	Ducts: Round Duct: Taps / Short Radius	Element ID: 1676269	<No level>	File > file.nwc > <No level> > Structural Framing > Structural1 Framing 650x19 > Structural1 Framing 650x400 > Structural1 Framing 650x19 > Structural1 Framing 650x400	Structural1 Framing 650x400	Structural Framing
	Clash5	Reviewed	-0.127		Hard	2018/12/2 07:26	x:-30.295, y:-0.111, z:7.563	Element ID: 532637	Ground Floor	File > file.nwc > Ground Floor > Ducts > Round Duct > Taps / Short Radius > Round Duct	Round Duct	Ducts: Round Duct: Taps / Short Radius	Element ID: 1619942	<No level>	File > file.nwc > <No level> > Structural Framing > Structural Framing 350x19 > Structural Framing 350x750 > Structural Framing 350x19 > Structural Framing 350x750	Structural Framing 350x750	Structural Framing
	Clash6	Reviewed	-0.127		Hard	2018/12/2 07:26	x:-30.295, y:0.589, z:7.563	Element ID: 526757	Ground Floor	File > file.nwc > Ground Floor > Ducts > Round Duct > Taps / Short Radius > Round Duct	Round Duct	Ducts: Round Duct: Taps / Short Radius	Element ID: 1619942	<No level>	File > file.nwc > <No level> > Structural Framing > Structural Framing 350x19 > Structural Framing 350x750 > Structural Framing 350x19 > Structural Framing 350x750	Structural Framing 350x750	Structural Framing

Figure 8-3 *Report created in the HTML format*

TUTORIALS

General instructions for downloading tutorial files:

1. Download the *c08_navisworks_2024_tut* zip file for this tutorial from *https://www.cadsofttech.com*. The path of the file is as follows: ***Books > Civil/GIS Books > Autodesk Navisworks > Exploring Autodesk Navisworks 2024 for Novices (Learn By Doing) > Tutorial Files.***

2. Now, save and extract the downloaded folder at the following location:

Note
The default unit system used in the tutorials is metric. To change the units to imperial, select the required units from ***Options Editor > Interface > Display Units****.*

Tutorial 1 Clash Detection Between Ducts and Beams

In this tutorial, you will perform clash detection to find the interferences between the ducts and beams in a model. To perform this tutorial, you will use the *c08_navisworks_2024_tut1.nwf* file.
(Expected time: 30 min)

The following steps are required to complete this tutorial:

a. Open the *c08_nws__tut1.nwf* file.
b. Create selection sets.
c. Add test and select test items.
d. Specify rules for clash detection.
e. Perform the clash test.
f. Review clash results.
g. Save the clashed views.
h. Create the clash report.
i. Save the project.

Opening the File

In this section, you will open the file.

1. Choose the **Open** button from the Quick Access Toolbar; the **Open** dialog box is displayed.

2. In this dialog box, browse to the following location:
   C:\nws_\c08_nws_2024_tut

3. Select the **Navisworks File Set (*.nwf)** file format from the **Files of type** drop-down list.

4. Next, select the *c08_nws_2024_tut1* from the file list; the file name is displayed in the **File name** drop-down.

5. Choose the **Open** button; the model is displayed, as shown in Figure 8-4.

Note

Sometimes on choosing the ***Open*** *button, the* ***Resolve*** *message box is displayed prompting you to resolve external references. In such a case, choose the* ***Ignore All*** *button to ignore the message.*

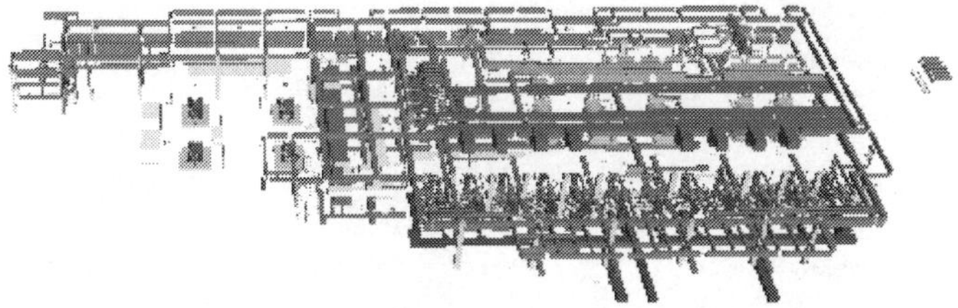

Figure 8-4 *Model displayed in the Scene View*

Creating Selection Sets

In this section, you will define two selection sets: Ducts and Beams. Further, you will use this selection set to perform the clash detection test.

1. Choose the **Selection Tree** tool from the **Select & Search** panel of the **Home** tab; the **Selection Tree** window is displayed.

2. Select the **Round Duct** sub node from **file.nwc > Ground Floor > Ducts** node, refer to Figure 8-5; the ducts are highlighted in the Scene View.

3. Choose the **Manage Sets** tool from **Home > Select & Search > Sets** drop-down; the **Sets** window is displayed.

4. In the **Sets** window, choose the **Save Selection** button; a selection set is created .

5. Enter the name **Ducts** in the selection set and press ENTER.

6. Next, select the **Structural Framing** sub-node in the **No Level** node of the **Selection Tree** window.

7. Repeat step 4 and save the selection with the name **Beams**. Figure 8-6 shows the **Sets** window with the created selection sets.

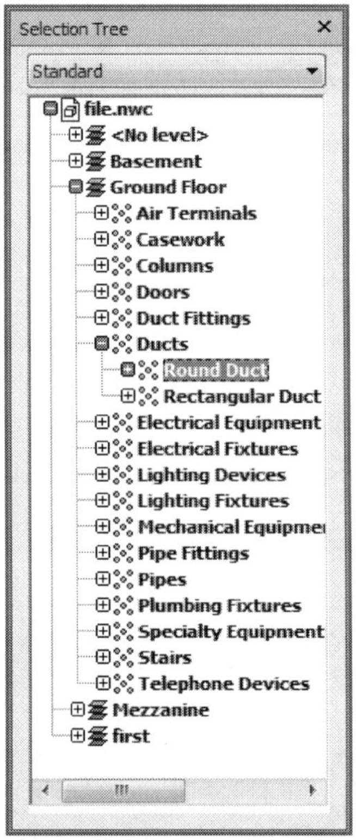

Figure 8-5 ***Round Duct*** *selected in the* ***Selection Tree*** *window*

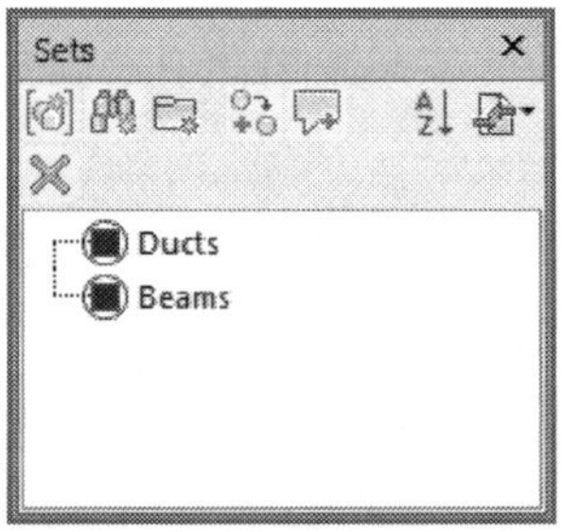

Figure 8-6 *The* ***Sets*** *window*

Adding the Test and Selecting Test Items

In this section, you will add a test in the **Clash Detective** window and then select the test items used for clash testing.

1. Choose the **Clash Detective** tool from the **Tools** panel in the **Home** tab; the **Clash Detective** window is displayed.

2. Choose the **Add Test** button in the **Clash Detective** window; a new test named **Test 1** is added in the **Test** panel.

3. Next, rename **Test1** as **Ducts vs Beams** by double-clicking on it and press ENTER, refer to Figure 8-7.

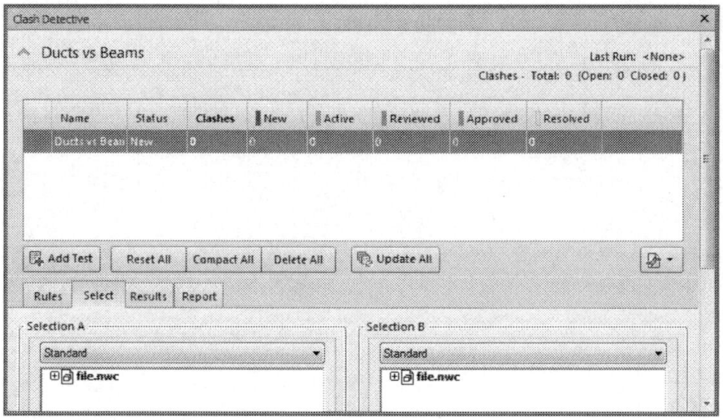

*Figure 8-7 Test added in the **Test** panel of the **Clash Detective** window*

4. Choose the **Select** tab in the **Clash Detective** window, if not chosen by default.

5. In the **Select** tab, select the **Sets** option from the drop-down lists available at the top in the **Selection A** and **Selection B** areas; the created selection sets are displayed in both the areas of the window.

6. Next, select **Ducts** in the **Selection A** area and **Beams** in the **Selection B** area, as shown in Figure 8-8.

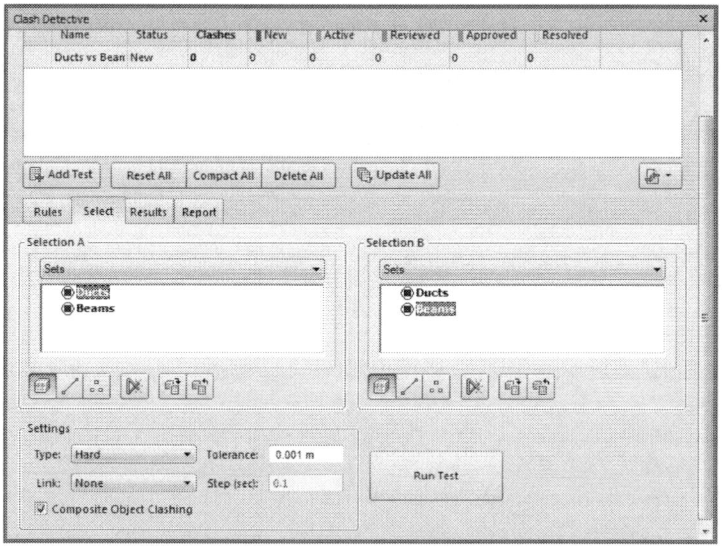

*Figure 8-8 Test items selected in the **Selection A** and **Selection B** areas*

Specifying Rules for Clash Detection

In this section, you will specify the rules to be used for detecting clashes.

1. Choose the **Rules** tab in the **Clash Detective** window.

2. Ensure that the **Items in same layer** check box is selected from the predefined rules displayed in the window, as shown in Figure 8-9.

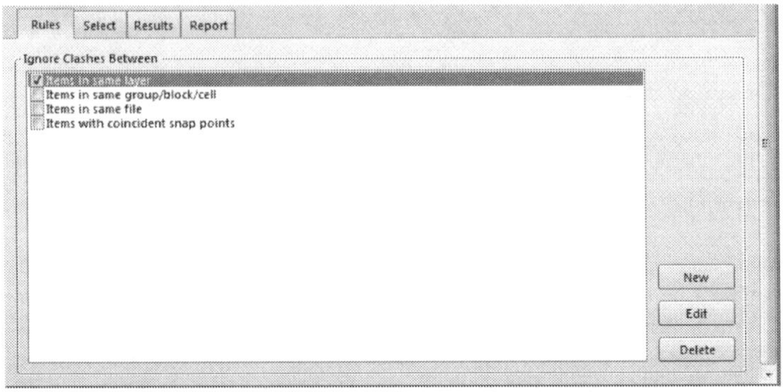

Figure 8-9 *Rule selected in the* ***Rules*** *tab*

Performing Clash Test

In this section, you will perform the clash test.

1. Choose the **Select** tab and ensure that the **Hard** option is selected in the **Type** drop-down in the **Settings** area.

2. Next, choose the **Run Test** button in the **Clash Detective** window; the **Results** tab is displayed with the clash results and the summary is displayed in the **Test** panel, as shown in Figure 8-10.

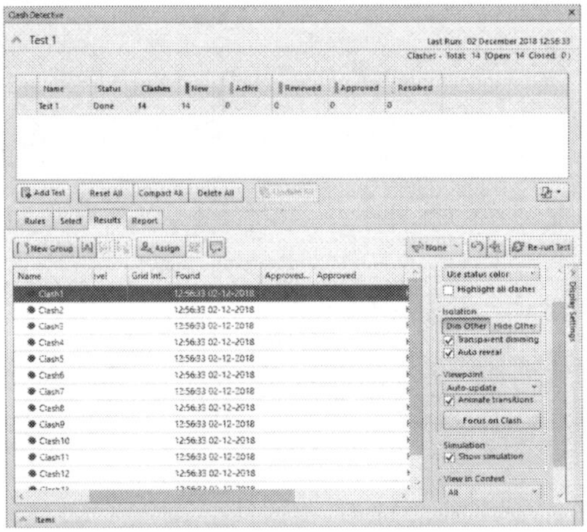

Figure 8-10 *Partial view of the clash test results displayed in the* ***Results*** *tab*

Reviewing Clash Results

In this section, you will view the clash results.

1. In the **Results** tab, ensure that the **Item 1** and **Item 2** buttons are chosen in the **Highlighting** area of the **Display Settings** panel.

2. Next, ensure that the **Dim Other** button is chosen in the **Isolation** area of the **Display Settings** panel.

3. Next, select the **Transparent dimming** and **Auto reveal** check boxes in the **Isolation** area.

4. Ensure that the **Auto-update** option is selected in the **Viewpoint** drop-down list.

5. Select the **Animate transitions** check box in the **Viewpoint** area of the **Display Settings** panel, if not selected.

Note

Deselect the clashing objects if selected in the Scene View.

6. Next, select the **Clash1** result in the result table; the corresponding clashed items are displayed in the Scene View, as shown in Figure 8-11.

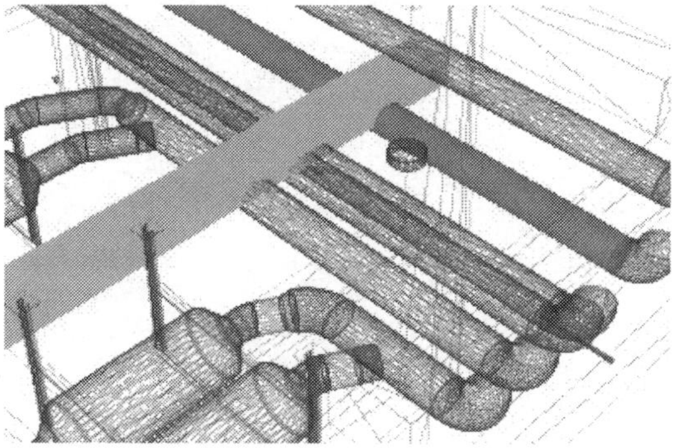

Figure 8-11 *Clashing objects in the Scene View*

7. Click in the **Status** column corresponding to the **Clash1** result; a drop-down list is displayed.

8. Select the **Reviewed** option from this list. Next, select the **Use status color** option from the drop-down list in the **Highlighting** area of the **Display Settings** panel; the clashed items get highlighted in the color assigned for the clash status, as shown in Figure 8-12.

9. Repeat the procedure followed in steps 1 to 7 and view the clash results from Clash2 to Clash14 clashes.

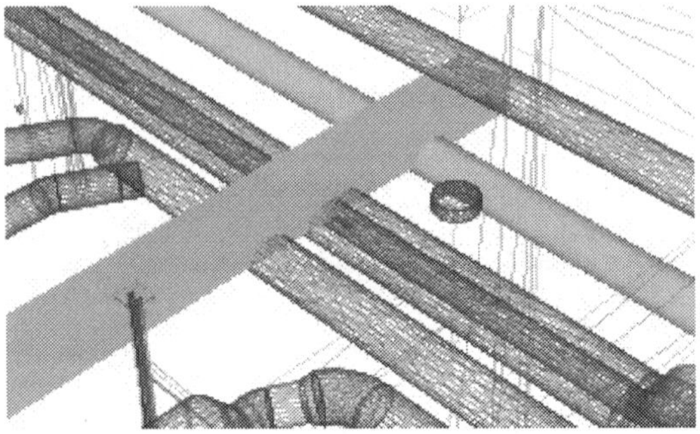

Figure 8-12 *Reviewed clash objects*

Saving Clashed Views

In this section, you will highlight the clashed item and add text to it.

1. Choose the **Cloud** tool from **Review > Markup > Draw** drop-down; the cursor changes into a pencil cursor.

2. Draw the cloud around the clashed items, as shown in Figure 8-13. Here, Clash1 has been shown.

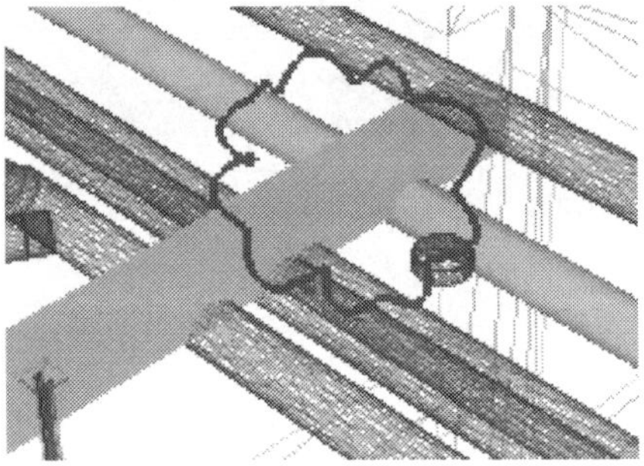

Figure 8-13 *Clouds drawn around the clashing objects*

3. Next, choose the **Text** tool from the **Markup** panel of the **Review** tab. Place the cursor inside the drawn cloud and click; the **Autodesk Navisworks Manage** dialog box is displayed.

4. In this dialog box, enter **Clashed Items** in the **Enter Markup Text** text box, and choose the **OK** button; the text is added, as shown in Figure 8-14.

5. Repeat the procedure followed in steps 1 to 4 and save the other clashed views.

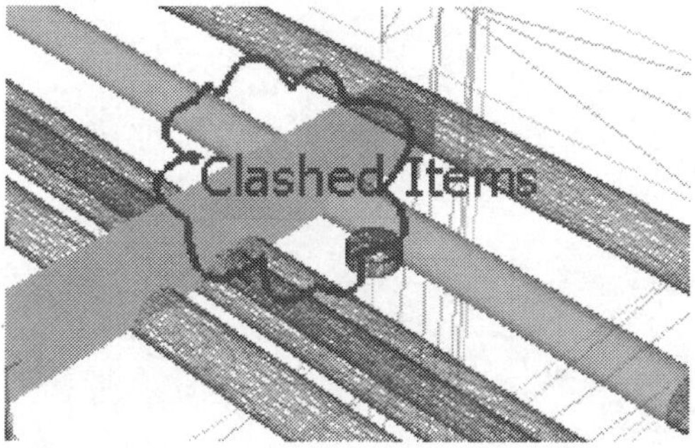

Figure 8-14 *Text added to the clashing objects*

Creating Clash Report

In this section, you will create the clash report.

1. Choose the **Report** tab in the **Clash Detective** window.

2. In the **Contents** and **Include these statuses** areas of the window, ensure that all the check boxes are selected, as shown in Figure 8-15.

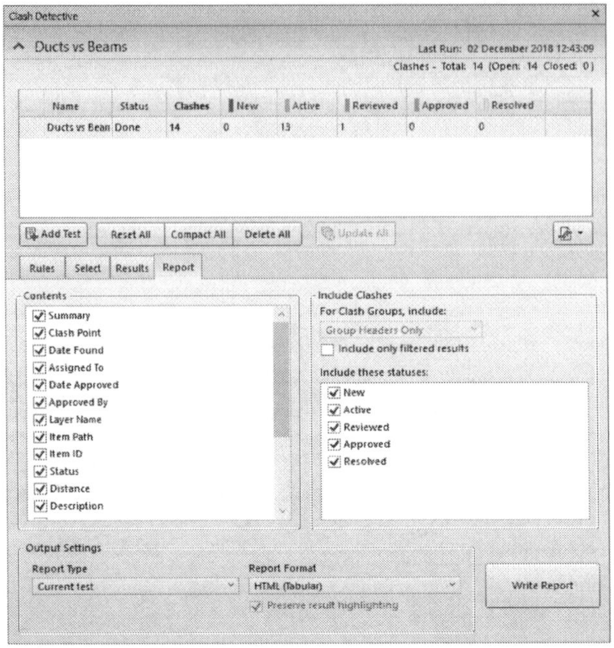

Figure 8-15 *Check boxes selected in the **Report** tab of the **Clash Detective** window*

3. Next, in the **Output Settings** area, ensure that the **Current test** option is selected in the **Report Type** drop-down list, refer to Figure 8-15.

4. Select the **HTML (Tabular)** option from the **Report Format** drop-down list, refer to Figure 8-15.

5. Next, choose the **Write Report** button; the **Save As** dialog box is displayed.

6. In this dialog box, browse to the following location: *C:\Navisworks_2024\c08_tutorial*.

7. Specify the name **Clash_Report_tut01** in the **File name** edit box and choose the **Save** button; the clash report is saved at the specified location.

8. Next, browse to the location on your system where the report is saved and click on the **Clash_Report_tut01** HTML document; the created report is displayed, as shown in Figure 8-16.

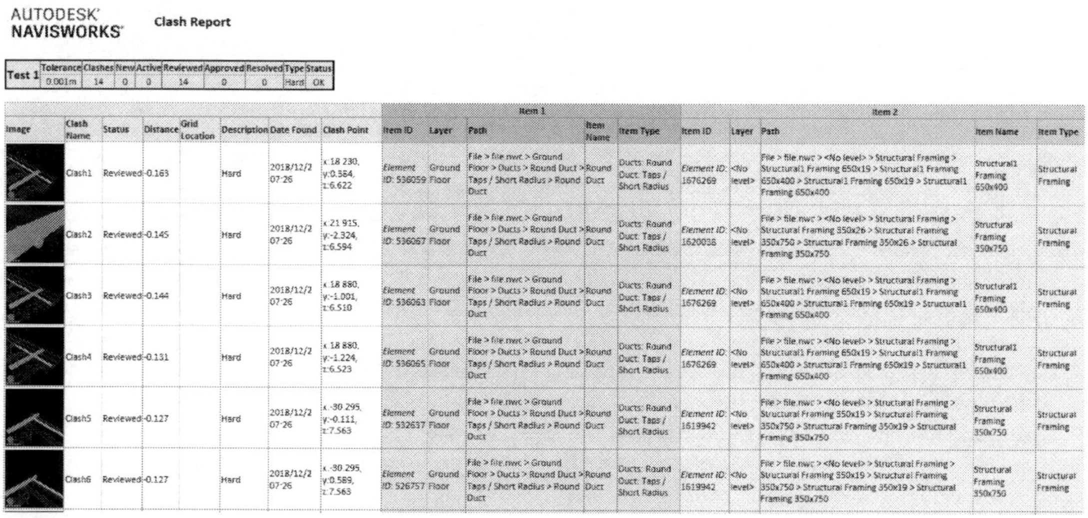
AUTODESK® NAVISWORKS® Clash Report

Test 1	Tolerance	Clashes	New	Active	Reviewed	Approved	Resolved	Type	Status
	0.001m	14	0	0	14	0	0	Hard	OK

								Item 1					Item 2				
Image	Clash Name	Status	Distance	Grid Location	Description	Date Found	Clash Point	Item ID	Layer	Path	Item Name	Item Type	Item ID	Layer	Path	Item Name	Item Type
	Clash1	Reviewed	-0.163		Hard	2018/12/2 07:26	x:18.230, y:0.364, z:6.622	Element ID: 536059	Ground Floor	File > file.nwc > Ground Floor > Ducts > Round Duct > Taps / Short Radius > Round Duct	Round Duct	Ducts: Round Duct: Taps / Short Radius	Element ID: 1676269	<No level>	File > file.nwc > <No level> > Structural Framing > Structural1 Framing 650x19 > Structural1 Framing 650x400 > Structural1 Framing 650x19 > Structural1 Framing 650x400	Structural1 Framing 650x400	Structural Framing
	Clash2	Reviewed	-0.145		Hard	2018/12/2 07:26	x:21.915, y:-2.324, z:6.594	Element ID: 536067	Ground Floor	File > file.nwc > Ground Floor > Ducts > Round Duct > Taps / Short Radius > Round Duct	Round Duct	Ducts: Round Duct: Taps / Short Radius	Element ID: 1620038	<No level>	File > file.nwc > <No level> > Structural Framing > Structural Framing 350x26 > Structural Framing 350x750 > Structural Framing 350x26 > Structural Framing 350x750	Structural Framing 350x750	Structural Framing
	Clash3	Reviewed	-0.144		Hard	2018/12/2 07:26	x:18.880, y:-1.001, z:6.510	Element ID: 536063	Ground Floor	File > file.nwc > Ground Floor > Ducts > Round Duct > Taps / Short Radius > Round Duct	Round Duct	Ducts: Round Duct: Taps / Short Radius	Element ID: 1676269	<No level>	File > file.nwc > <No level> > Structural Framing > Structural1 Framing 650x19 > Structural1 Framing 650x400 > Structural1 Framing 650x19 > Structural1 Framing 650x400	Structural1 Framing 650x400	Structural Framing
	Clash4	Reviewed	-0.131		Hard	2018/12/2 07:26	x:18.880, y:-1.224, z:6.523	Element ID: 536065	Ground Floor	File > file.nwc > Ground Floor > Ducts > Round Duct > Taps / Short Radius > Round Duct	Round Duct	Ducts: Round Duct: Taps / Short Radius	Element ID: 1676269	<No level>	File > file.nwc > <No level> > Structural Framing > Structural1 Framing 650x19 > Structural1 Framing 650x400 > Structural1 Framing 650x19 > Structural1 Framing 650x400	Structural1 Framing 650x400	Structural Framing
	Clash5	Reviewed	-0.127		Hard	2018/12/2 07:26	x:-30.295, y:-0.111, z:7.563	Element ID: 532637	Ground Floor	File > file.nwc > Ground Floor > Ducts > Round Duct > Taps / Short Radius > Round Duct	Round Duct	Ducts: Round Duct: Taps / Short Radius	Element ID: 1619942	<No level>	File > file.nwc > <No level> > Structural Framing > Structural Framing 350x19 > Structural Framing 350x750 > Structural Framing 350x19 > Structural Framing 350x750	Structural Framing 350x750	Structural Framing
	Clash6	Reviewed	-0.127		Hard	2018/12/2 07:26	x:-30.295, y:0.589, z:7.563	Element ID: 526757	Ground Floor	File > file.nwc > Ground Floor > Ducts > Round Duct > Taps / Short Radius > Round Duct	Round Duct	Ducts: Round Duct: Taps / Short Radius	Element ID: 1619942	<No level>	File > file.nwc > <No level> > Structural Framing > Structural Framing 350x19 > Structural Framing 350x750 > Structural Framing 350x19 > Structural Framing 350x750	Structural Framing 350x750	Structural Framing

***Figure 8-16** The clash report generated in the HTML Tabular format*

Saving the Project

In this section, you will save the project.

1. Close the **Clash Detective** window.

2. Select the **3D View** in the **Saved Viewpoints** window; the view is displayed in the Scene View.

3. Choose the **Save As** option from the Application Menu, the **Save As** dialog box is displayed.

4. Browse to *nws_2024_tut / nws_2024_tut* folder and enter *c08_navisworks_2024_tut01* in the **File name** edit box. Next, choose **Navisworks File Set (*.nwf)** file format from the **Save as type** drop-down list, and choose the **Save** button; the project is saved.

Tutorial 2 Clash Detection Between Pipes and Walls

In this tutorial, you will perform clash detection to find the interferences between pipes and walls in a model. To perform this tutorial, you will use the *c08_navisworks_2024_tut2* file.

(Expected time: 20 min)

The following steps are required to complete this tutorial:

a. Open the *c08_navisworks_2024_tut2* file.
b. Add the test and select the test items.
c. Perform the clash test.
d. Review clash results.
e. Save the clashed views.
f. Create clash report.
g. Save the project.

Opening the File

In this section, you will open the file.

1. Choose the **Open** button from the Quick Access Toolbar; the **Open** dialog box is displayed on the screen.

2. In this dialog box, browse to the following location:
   C:\nws_2024\c08_nws_2024_tut

3. Select the **Navisworks File Set (*.nwf)** file format from the **Files of type** drop-down list.

4. Next, select the *c08_navisworks_2024_tut2* from the displayed list of files; the file name appears in the **File name** edit box.

5. Choose the **Open** button; the model is displayed, as shown in Figure 8-17.

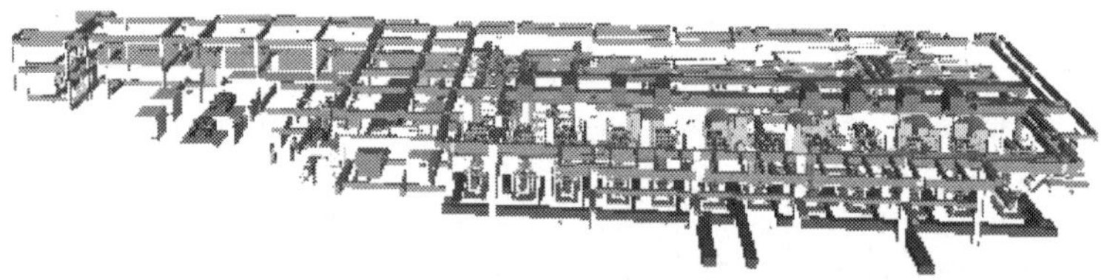

Figure 8-17 *Model displayed in the Scene View*

Adding the Test and Selecting Test Items

In this section, you will add a test to the **Clash Detective** window and then select the test items used for clash testing.

1. Choose the **Clash Detective** tool from the **Tools** panel in the **Home** tab; the **Clash Detective** window is displayed.

   Alternatively, you can press the **CTRL + F2** buttons to invoke the **Clash Detective** window.

2. Choose the **Add Test** button in the **Clash Detective** window; **Test 1** is added to the **Test** panel.

3. Rename **Test1** as **Pipes vs Walls** by double-clicking on it and press ENTER, refer to Figure 8-18.

4. Choose the **Select** tab in the **Clash Detective** window.

5. In the **Select** tab, select the **Sets** option from the drop-down lists available at the top in the **Selection A** and **Selection B** areas; the created selection sets will be displayed in both the areas of the window.

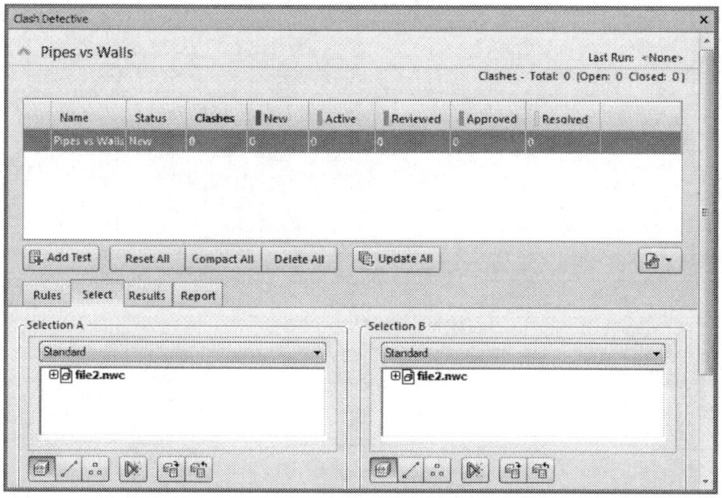

Figure 8-18 Test added to the ***Test*** *panel of the* ***Clash Detective*** *window*

6. Select **Pipes** in the **Selection A** area and **Walls** in the **Selection B** areas, refer to Figure 8-19.

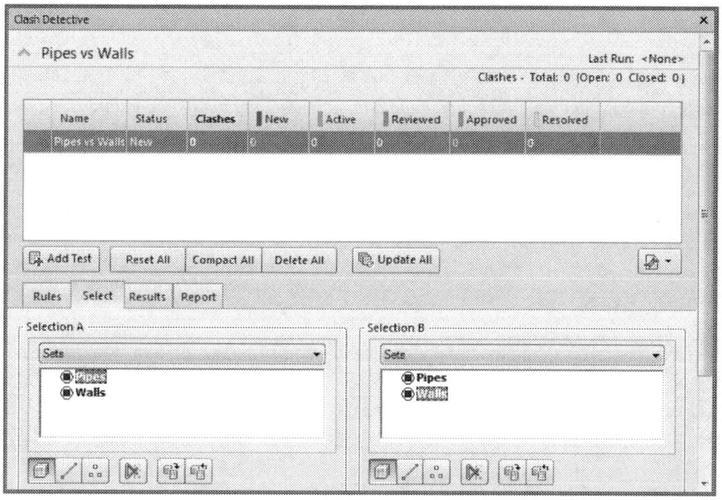

Figure 8-19 Test items selected in the ***Selection A*** *and* ***Selection B*** *areas*

Note
Ensure that all check boxes in the ***Ignore Clashes Between*** *area of the* ***Rules*** *tab are cleared.*

Performing the Clash Test

In this section, you will perform clash test between various building components.

1. In the **Select** tab, ensure that the **Hard** option is selected in the **Type** drop-down list in the **Settings** area.

2. Choose the **Run Test** button in the **Clash Detective** window; the **Results** tab is displayed with all the clash results. Also, the summary is displayed in the **Test** panel, as shown in Figure 8-20.

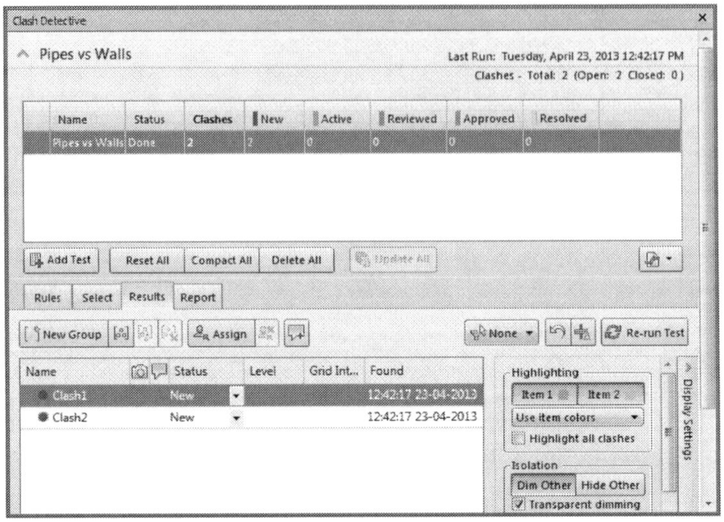

Figure 8-20 *Clash results and summary displayed in the window*

Reviewing Clash Results

In this section, you will view clash results.

1. In the **Results** tab, ensure that **Item 1** and **Item 2** buttons are chosen in the **Highlighting** area of the **Display Settings** panel.

2. Next, ensure that the **Dim Other** button is chosen in the **Isolation** area of the **Display Settings** panel.

3. Next, select the **Transparent dimming** and **Auto reveal** check boxes in the **Isolation** area, if it is not selected by default.

4. Ensure that the **Auto-update** option is selected in the **Viewpoint** drop-down list.

5. Select the **Animate transitions** check box, if it is not selected by default.

6. Next, select the **Clash1** result in the result table; the corresponding clashed items are displayed in the Scene View, as shown in Figure 8-21.

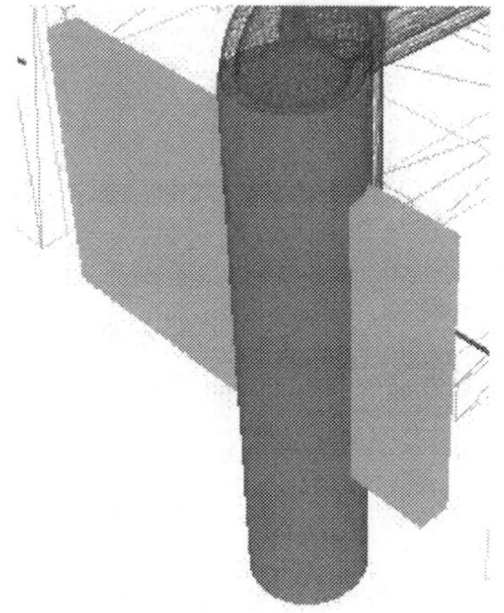

Figure 8-21 Clash objects in the Scene View

7. Click in the **Status** column corresponding to the **Clash1** result; a drop-down list is displayed.

8. Select the **Reviewed** option from this list. Next, select the **Use status color** option from the drop-down list in the **Highlighting** area of the **Display Settings** panel; the clashed items are highlighted with the color assigned for the clash status, as shown in Figure 8-22.

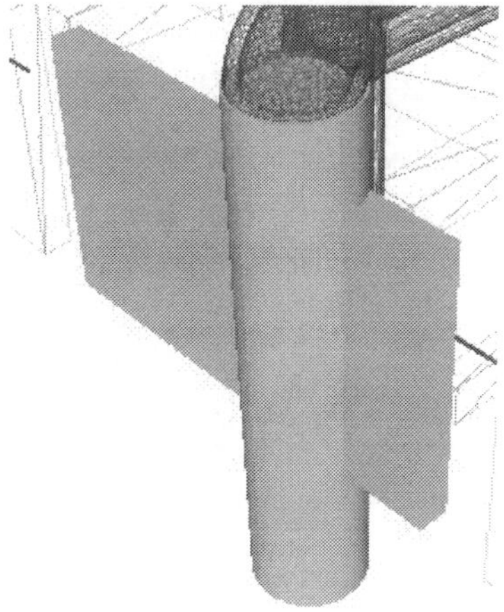

Figure 8-22 Reviewed clash objects in the Scene View

9. Repeat the procedure followed in step 7 and approve the **Clash2** result by selecting the **Approved** option from the **Status** drop-down list.

Saving the Clashed Views

In this section, you will highlight the clashed item and add texts to them.

1. Select the **Clash1** result in the **Results** tab; the clash1 will be visible in the scene view. Choose the **Ellipse** tool from **Review > Markup > Draw** drop-down; the cursor changes into a pencil cursor.

2. Draw the ellipse around the clashed objects of the **Clash1** result, as shown in Figure 8-23.

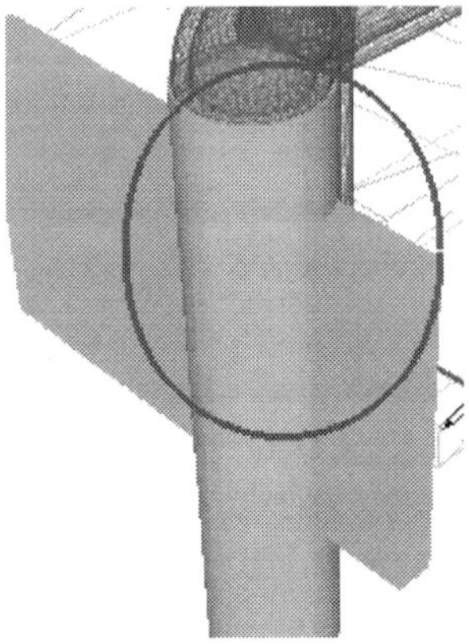

Figure 8-23 Ellipse drawn around the clashed objects

3. Next, choose the **Text** tool from the **Markup** panel of the **Review** tab. Place the cursor inside the drawn ellipse and click; the **Autodesk Navisworks Manage** dialog box is displayed.

4. In this dialog box, enter the text **Clashed Objects** in the **Enter Markup Text** text box, and choose the **OK** button; the text is added, as shown in Figure 8-24.

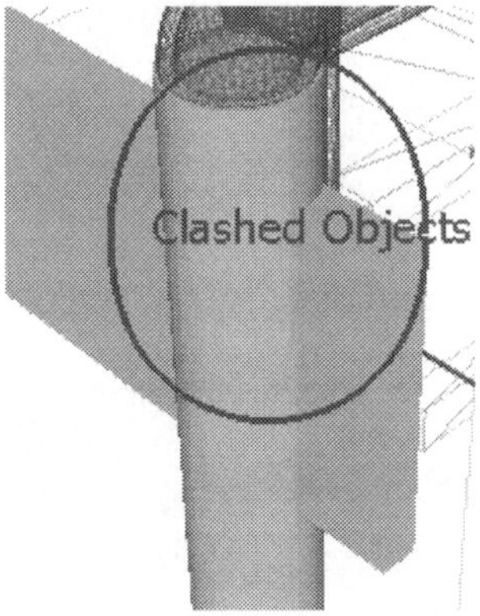

Figure 8-24 Text added to the clashing objects

5. Repeat the procedure followed in steps 1 to 4 and save the **Clash2** result view.

Creating Clash Report

In this section, you will create the clash report.

1. Choose the **Report** tab in the **Clash Detective** window.

2. In the **Contents** and **Include these statuses** areas, select the check boxes, as shown in Figure 8-25.

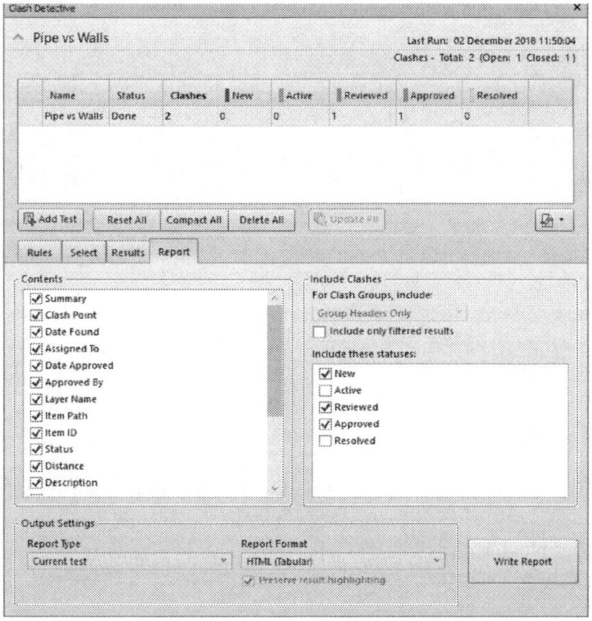

*Figure 8-25 The **Report** tab in the **Clash Detective** window*

3. Next, in the **Output Settings** area, ensure that the **Current test** option is selected in the **Report Type** drop-down list, refer to Figure 8-25.

4. Select the **Text** option from the **Report Format** drop-down list, refer to Figure 8-38.

5. Next, choose the **Write Report** button; the **Save As** dialog box is displayed.

6. In this dialog box, browse to the following location: *C:\nws_2024\c08_nws_2024_tut*.

7. Enter the name **Clash_Report_tut02** in the **File name** edit box and choose the **Save** button; the clash report is saved at the specified location.

8. Next, browse to the location on your system where the report is saved and click on the **Clash_Report_tut02** text document; the created report is displayed in the text format, as shown in Figure 8-26.

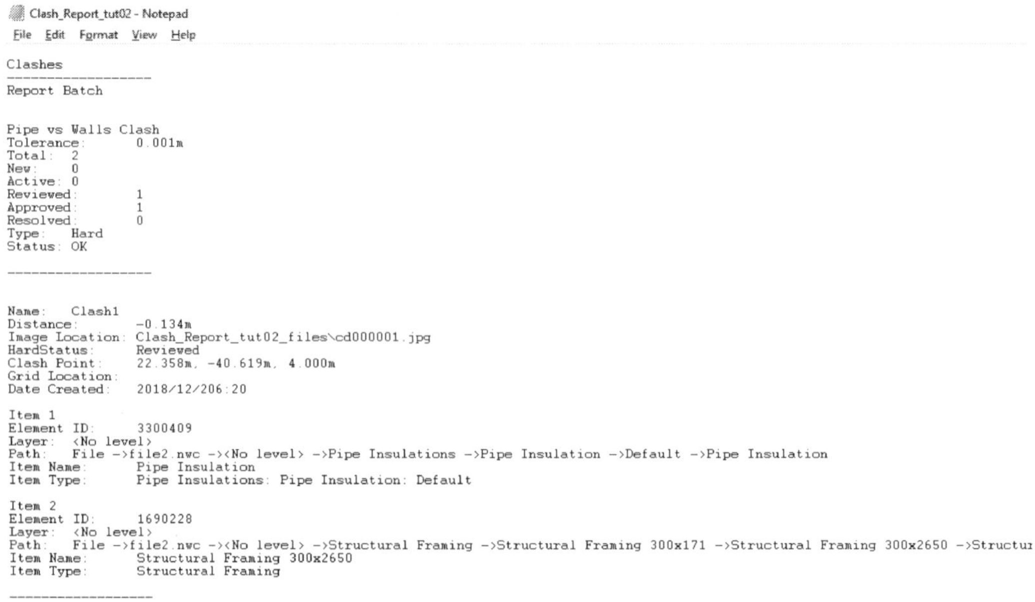

```
Clash_Report_tut02 - Notepad
File  Edit  Format  View  Help

Clashes
------------------
Report Batch


Pipe vs Walls Clash
Tolerance:      0.001m
Total:  2
New:    0
Active: 0
Reviewed:       1
Approved:       1
Resolved:       0
Type:   Hard
Status: OK

------------------


Name:   Clash1
Distance:       -0.134m
Image Location: Clash_Report_tut02_files\cd000001.jpg
HardStatus:     Reviewed
Clash Point:    22.358m, -40.619m, 4.000m
Grid Location:
Date Created:   2018/12/206:20

Item 1
Element ID:     3300409
Layer:  <No level>
Path:   File ->file2.nwc -><No level> ->Pipe Insulations ->Pipe Insulation ->Default ->Pipe Insulation
Item Name:      Pipe Insulation
Item Type:      Pipe Insulations: Pipe Insulation: Default

Item 2
Element ID:     1690228
Layer:  <No level>
Path:   File ->file2.nwc -><No level> ->Structural Framing ->Structural Framing 300x171 ->Structural Framing 300x2650 ->Structur
Item Name:      Structural Framing 300x2650
Item Type:      Structural Framing

------------------
```

***Figure 8-26** Partial view of the clash report in the text format*

Saving the Project

In this section, you will save the project.

1. Close the **Clash Detective** window.

2. Select the **3D View** in the **Saved Viewpoints** window; the view is displayed in the Scene View.

3. Choose the **Save As** option from the Application Menu; the **Save As** dialog box is displayed.

4. Browse to *nws_2024\ c08_nws_2024_tut* folder and enter *c08_navisworks_2024_tut02* in the **File name** edit box. Next, choose **Navisworks File Set (*.nwf)** file format from the **Save as type** drop-down list, and choose the **Save** button.

SELF-EVALUATION TEST

Answer the following questions and then compare them to those given at the end of this chapter:

1. You can reset the status of all the clash tests to **New** by choosing the __________ button.

2. You can use the __________ rule to ignore clashes between objects in the same layer.

3. You can include the surface geometries in the clash tests by choosing the __________ button.

4. The __________ button is used to ungroup the selected clash results.

5. You can fade the objects that are not involved in the clash test by choosing the _________ button.

6. You cannot link a clash test to an object animation. (T/F)

7. You cannot include point geometry in a clash test. (T/F)

8. You cannot customize the rules available in the **Rules** tab. (T/F)

9. The **Test** panel displays the summary of the performed clash test. (T/F)

10. You cannot ignore clashes between objects of the same selection set. (T/F)

REVIEW QUESTIONS

Answer the following questions:

1. Which of the following options is used to remove all the resolved clash tests?

   a) **Update All** b) **Delete All**
   c) **Compact All** d) **Reset All**

2. Which of the following options is used to include line geometry in a clash test?

   a) **Point** b) **Surfaces**
   c) **Self-Intersect** d) **Lines**

3. Which of the following options is used to perform a clash test?

   a) **Start** b) **Play**
   c) **Run Test** d) **Stop**

4. Which of the following options is used to group the selected clash results?

   a) **Remove from Group** b) **New Group**
   c) **Group Selected Clashes** d) **Explode Group**

5. Which of the following options is used to assign a clash or a clash group to a person or trade?

   a) **Unassign** b) **Assign**
   c) **Delete** d) **Reset All**

6. You cannot save the viewpoint for the clash results. (T/F)

7. You can highlight the objects involved in a clash by choosing the **Select in Scene** button. (T/F)

8. The **Unassign** button is used to clear the assignment for selected clashes. (T/F)

9. You cannot add comments to the clashes. (T/F)

10. You cannot customize the display of clash results in the Scene View. (T/F)

EXERCISE

Exercise 1 Performing Clash Detection

Download the *c08_navisworks_2024_ex1.nwf* file from *https://www.cadsofttech.com* and then open it. Find clashes between model elements using the **Clash Detective** window. Figure 8-27 shows the model to be used for the clash detection. Use the parameters given below to complete this exercise. **(Expected time: 45 min)**

Project parameters for Exercise 1:

Clash Test 1: **Ducts vs Walls**
Clash Test 2: **Pipes vs Walls**
Report format: **HTML (Tabular)**

The following steps are required to complete this exercise:

a. Open the model and display the **Clash Detective** window.
b. Add a test and select objects for the clash test.
c. Use the selection set defined in the **Sets** window for the clash detection.
d. Run the test and view the clash results.
e. Group clash results that are related to a single issue.
f. Assign the clash status as reviewed in both the clash tests.
g. Create a combined clash report with the name *Clash_Report_ex01*.
h. Save the project as *c08_navisworks_2024_ex01*.

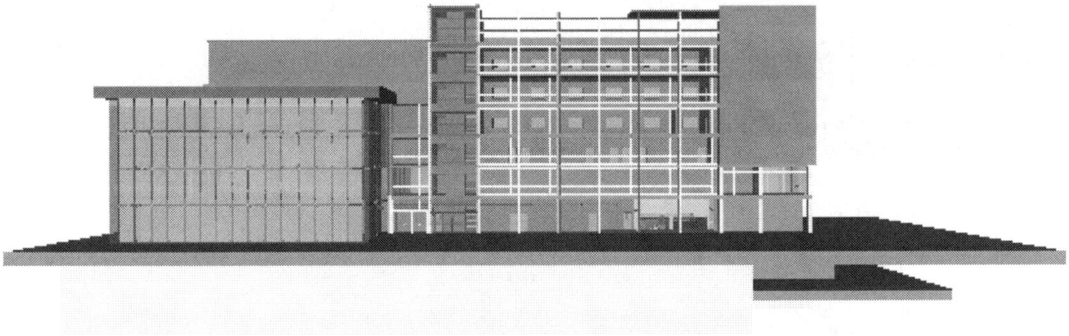

Figure 8-27 *The Conference Center*

Answers to Self-Evaluation Test

1. Reset All, **2. Items in Same Layer**, **3. Surfaces**, **4. Explode Group**, **5. Dim Other**, **6.** F, **7.** F, **8.** F, **9.** T, **10.** F

Chapter 9

Autodesk Rendering in Navisworks

Learning Objectives

After completing this chapter, you will be able to:

- *Use Autodesk materials*
- *Use Autodesk lights*
- *Apply Sun, Sky, and Exposure effects*
- *Render using Autodesk Graphics*

INTRODUCTION

Autodesk Rendering feature in Navisworks allows you to achieve better photorealistic results with the help of the **Ray Trace** tool. In the language of computer graphics, ray tracing is a technique used for generating photorealistic image. This technique of rendering is capable of producing a visual realism much better than produced by typical scanline rendering methods but at a greater computational cost.

One of the advantages of using this Autodesk Rendering feature is that on opening a model created in any of the Autodesk products, all the materials assigned to the model will be displayed in Navisworks in a rendered view. You do not need to apply the materials again to the model in Navisworks.

Autodesk Rendering feature allows you to use existing material libraries, apply lightings, and adjust environment settings. You can also create custom libraries for a specific project using this feature.

THE AUTODESK RENDERING WINDOW

In Navisworks, the **Autodesk Rendering** window can be used to apply Autodesk material libraries and lightings. Also, you can change the environment settings of a project using this window as shown in Figure 9-1.

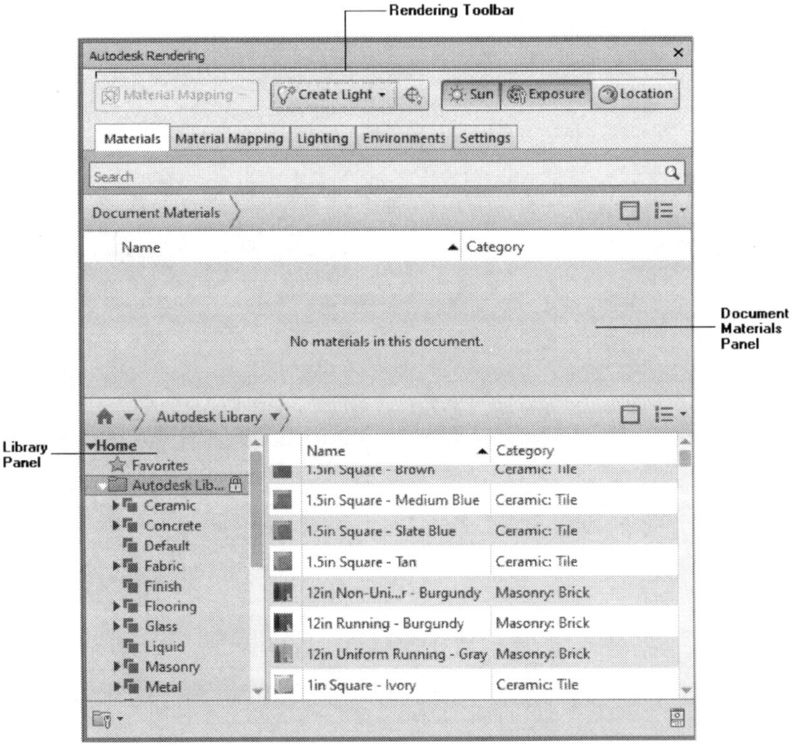

*Figure 9-1 The **Autodesk Rendering** window*

The **Autodesk Rendering** window comprises of the Rendering toolbar and five tabs: **Materials**, **Material Mapping**, **Lighting**, **Environments**, and **Settings**.

The Rendering Toolbar

The Rendering toolbar is located at the top of the **Autodesk Rendering** window. It contains various drop-down lists and buttons that are used for material mapping, creating and placing lights, adjusting exposure settings, and specifying geographical location of the model.

Materials Tab

In the **Materials** tab, you can manage materials, apply materials to the selected objects in the model, and can create custom materials. The **Materials** tab comprises the Search area, Library panel, and the **Document Materials** panel, refer to Figure 9-1.

Material Mapping Tab

In the **Material Mapping** tab, you can customize the settings of the material mapping type that is selected from the **Material Mapping** drop-down list. Note that the **Material Mapping** tab will not display parameters unless you select individual objects, such as polygon or an instanced item of geometry which are represented by and symbols in the **Selection Tree** window. When you select an object in the **Selection Tree** window, the material mapping parameters will be displayed according to the selected mapping type in the tab.

Lighting Tab

In the **Lighting** tab, you can manage lights in a model. The **Lighting** tab is divided into two areas, the Lights view and the Properties view.

The Lights view area contains a list of lights used in the scene, with their name and type. You can turn the lights on and off by selecting the **Status** check box in the Lights view area.

The Properties view area shows the properties of the selected light. When you select a light in the Lights view area, its properties will be displayed in the Properties view area. Using these properties, you can change the color of light, its intensity, and so on.

The lights in the Lights view area will be displayed only after adding lights to the scene.

Environments Tab

The options in the **Environments** tab are used to adjust the sun, sky, and exposure effects.

Settings Tab

In the **Settings** tab, various rendering styles and quality options are available which can be used torender the model.There are three areas in this tab: Current Render Preset, Basic, and Advanced.

RENDERING USING AUTODESK GRAPHICS

Using Autodesk Graphics, you can produce highly detailed and photorealistic images. After selecting the desired render style type in the **Autodesk Rendering** window, choose the **Ray Trace** button from the **Interactive Ray Trace** panel of the **Render** tab; the rendering process will start and you can see the render progress indicator on the screen. While rendering, you can pan, orbit, and zoom the model. However, performing these actions will restart the rendering process. You can control the rendering process by using the buttons available in the **Interactive Ray Trace** panel of the **Render** tab. Choose the **Pause** button to temporarily stop the rendering process. Choose the **Close** button to exit the rendering process. You can also save the rendered scene as an image. To do so, choose the **Save** button from the **Interactive Ray Trace** panel of the **Render** tab; the **Save As** dialog box will be displayed. In this dialog box, select the desired file type from the **Save As Type** drop-down list. Browse to the desired location to save the file. Enter the file name and choose the **Save** button; the image will be saved at the specified location.

RENDERING IN CLOUD

Autodesk has added a new capability of rendering in cloud that helps to render views online. You can use this application to access saved rendered images, change the render quality of existing rendered images, and use different background effects using the Autodesk render gallery.

TUTORIALS

General instructions for downloading tutorial files:

1. Download the *c09_navisworks_2024_tut* zip file for the tutorial from *www.cadsofttech.com*. The path of the file is as follows: ***Books > Civil/GIS Books > Autodesk Navisworks > Exploring Autodesk Navisworks 2024 for Novices (Learn By Doing) > Tutorial Files.***

2. Now, save and extract the downloaded folder at the following location:
   C:\ nws_2024\ c09_nws_2024_tut

Tutorial 1 — Rendering the Room

In this tutorial, you will open the *c09_navisworks_2024_tut1.nwf* file. Next, you will apply Autodesk materials and lights to the model elements such as furniture, walls, windows, and so on. After applying the materials and lights, you will render the scene using high quality rendering.

(Expected time: 45 min)

The following steps are required to complete this tutorial:

a. Open the file *c09_navisworks_2024_tut1.nwf.*
b. Invoke the **Autodesk Rendering** window.
c. Add materials to the **Document Materials** panel in the window.
d. Display the **Sets** window.
e. Apply materials to the model elements.
f. Add lights to the scene.
g. Specify the location, date, and time.
h. Render the model.
i. Save the project.

Opening the File

In this section, you will open the file in Navisworks.

1. Choose the **Open** button from the Quick Access Toolbar; the **Open** dialog box is displayed.

2. In this dialog box, browse to the following location:
   C:\nws_2024\c09_nws_2024_tut

3. Select **Navisworks File Set (*.nwf)** from the **Files of type** drop-down list.

4. Select the file *c09_navisworks_2024_tut1* from the displayed list of files; the file name appears in the **File name** edit box.

5. Choose the **Open** button in this dialog box; the model is displayed in the Scene View, as shown in Figure 9-2.

Figure 9-2 *The interior view of the model*

Invoking the Autodesk Rendering Window

In this section, you will invoke the **Autodesk Rendering** window.

1. Choose the **Autodesk Rendering** tool from the drop-down in the **Tools** panel of the **Home** tab; the **Autodesk Rendering** window is displayed. Alternatively, you can choose CTRL + F4 buttons to invoke the **Autodesk Rendering** window.

Note

*If materials are displayed in the **Document Materials** panel of this window, then delete them to add new materials.*

Adding Materials to the Document Materials Panel

In this section, you will add materials to the **Document Materials** panel in the **Autodesk Rendering** window.

1. In the **Autodesk Rendering** window, ensure that the **Materials** tab is chosen. In this tab, expand the **Autodesk Library** node in the Library panel if not expanded by default; various material folders are displayed.

2. Next, select the **Glossy** category under the **Autodesk Library > Wall Paint** node; various materials are displayed in the right pane of the Library panel.

3. In the right pane of the Library panel, scroll down and place the cursor on the **Maroon** wall paint; two buttons are displayed, as shown in Figure 9-3.

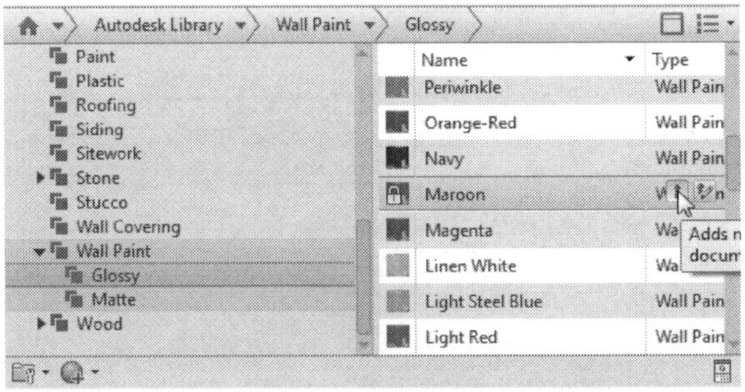

***Figure 9-3** Buttons displayed on selecting the wall paint*

4. Next, choose the **Adds material to document** button; the **Maroon** wall paint is added to the **Document Materials** panel of the **Autodesk Rendering** window, as shown in Figure 9-4.

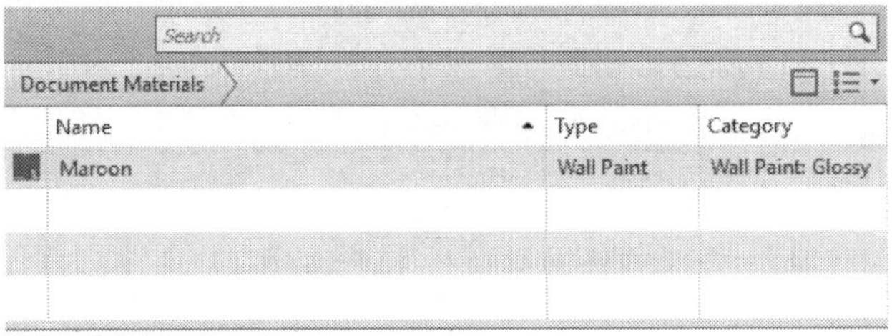

***Figure 9-4** The **Maroon** wall paint added to the **Document Materials** panel*

Next, you will add the materials to be used for window frames and glasses.

5. Expand the category under the **Autodesk Library > Wood** node; various materials are displayed in the right pane of the Library panel.

6. Select the **Teak - Stained Dark Medium Gloss** wood type and add it to the **Document Materials** panel, as explained in step 4. Similarly, add the **Teak - Natural Polished** wood type to the **Document Materials** panel.

7. Next, select the **Glass** node from the left pane of the Library panel; various glass types are displayed in the right pane.

8. Select the **Tinted - Blue** glass type and add it to the **Document Materials** panel.

9. Next, select **Glass > Glazing** from the left pane; different glass types are displayed in the right pane.

10. Select **Light Blue Reflective** glass type and add it to the **Document Materials** panel.

Repeat the procedure followed in the previous steps and add the fabric, carpet texture, and material for side table, lamps, and other fixtures respectively, as given in Table 9-1. Figure 9-5 shows the partial list of the materials added to the **Document Materials** panel.

Table 9-1 *List of materials to be added to the* ***Document Materials*** *panel*

Materials to be added	Path of the materials in the Library panel
Pebbled - Mauve	Autodesk Library > Fabric > Leather
Rug - Braided	Autodesk Library > Flooring > Carpet
Wax - Black on Red Wood	Autodesk Library > Finish
Wax - Red on Light Wood	Autodesk Library > Finish
Titanium - Polished	Autodesk Library > Metal
Chrome - Satin	Autodesk Library > Metal

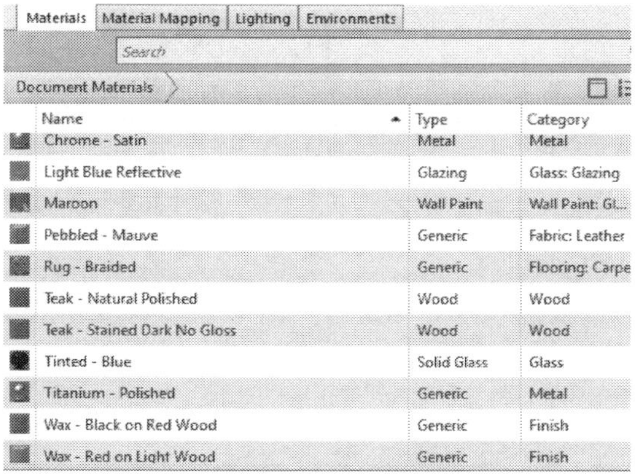

Figure 9-5 *Materials added to the* ***Document Materials*** *panel*

Displaying the Sets Window

In this section, you will display the **Sets** window.

1. Choose the **Manage Sets** tool from the **Home > Select & Search > Sets** drop-down; the **Sets** window is displayed, as shown in Figure 9-5.

In the **Sets** window, the selection sets are saved so that it is easier for you to apply the materials to the required model elements.

Applying Materials to the Model Elements

In this section, you will apply the added materials to the model elements such as windows, sofa, table, lamp, and so on.

1. In the **Sets** window, select **Wall**; all walls are highlighted in the Scene View.

2. Next, right-click on the **Maroon** wall paint in the **Document Materials** panel; a shortcut menu is displayed.

3. Choose the **Assign to Selection** option from the menu; the **Maroon** wall paint is applied on the walls.

4. Repeat the procedure followed in steps 1 to 3 and apply the materials to the model elements as given in Table 9-2. Figure 9-6 shows the model after applying all the materials.

Table 9-2 *List of objects and materials assigned to them*

Objects in the Sets window	Materials to be Assigned
Window Frame	Teak - Stained Dark Medium Gloss
Window Glass	Light Blue Reflective
Table Frame	Wax Red on Light Wood
Table Glass	Light Blue Reflective
Dining Table	Teak - Natural Polished
Side Table	Wax Black on Red Wood
Lamp	Chrome - Satin
Sofa	Pebbled - Mauve
Floor	Rug - Braided
Pendant Light	Titanium - Polished

Figure 9-6 View after applying all the materials

Adding Lights to the Scene

In this section, you will add lights in the scene.

1. In the **Autodesk Rendering** window, select the **Spot** light option from the **Create Light** drop-down list in the Rendering toolbar; the **Lighting** tab is displayed in the window. Also, a torch is attached to the cursor.
   Now, you will specify the location and target of the light.

2. Click on the lamp, as shown in Figure 9-7; the location of the Spot light is specified.

Note
*If there are **Point** lights in the **Light View** area, then delete them.*

Figure 9-7 Specifying the location for the Spot light

3. Move the cursor such that the torch points downward. Next, click in the drawing; the base of the Spot light is specified as the target. Note that the light gizmo is also displayed in the Scene View, as shown in Figure 9-8.

***Figure 9-8** Light gizmo displayed in the Scene View*

4. Next, specify the properties for the Spot light in the **Properties** view area, as shown in Figure 9-9.

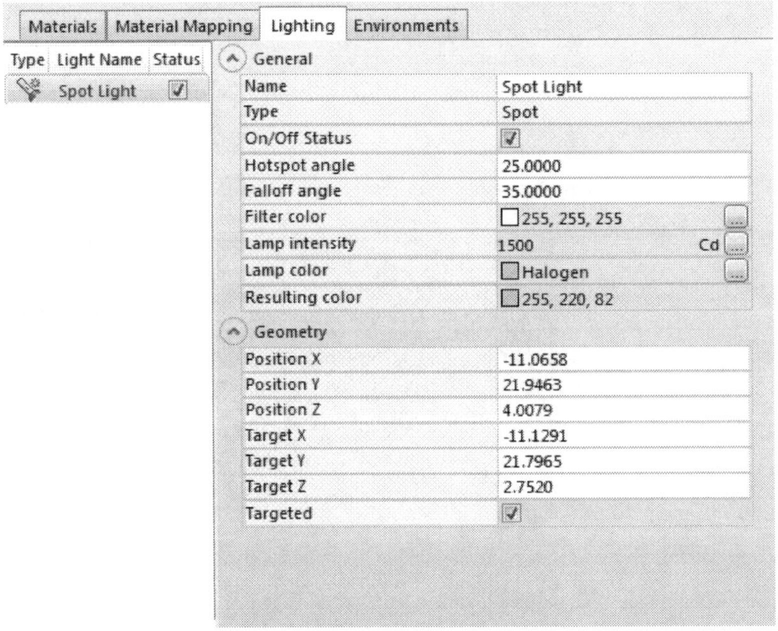

***Figure 9-9** Specifying parameters for the Spot light*

5. In the Properties view area of the **Lighting** tab, choose the browse button displayed in the field corresponding to the **Lamp Intensity** option in the **General** area; the **Lamp Intensity** dialog box is displayed.

6. In this dialog box, ensure that the **Intensity (Candela)** option is selected in the **Unit** drop-down list and enter **50** in the **Intensity** edit box.

7. Choose the **OK** button; the dialog box closes and the light intensity of the Spot light is changed accordingly.

Next, you will add point light to the scene.

8. Select the **Point** light option from the **Create Light** drop-down list in the Rendering toolbar; a light glyph is displayed along with the cursor.

9. Click to specify the position of the Point light in the Scene View, as shown in Figure 9-10; a gizmo is displayed in the Scene View. Also, the Point light is added to the **Properties** view area of the **Lighting** tab.

> **Tip**
> *Ensure that the **Full Lights** option is selected from the **Viewpoint > Render Style > Lighting** drop-down list. It will produce a high quality lighting.*

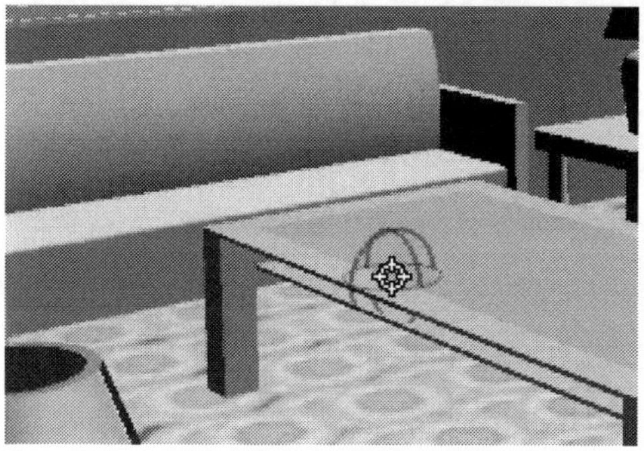

Figure 9-10 *Specifying location of light*

10. Specify the properties for the Point light in the **Properties** view area, as shown in Figure 9-11. Next, you will change the color of the Point light.

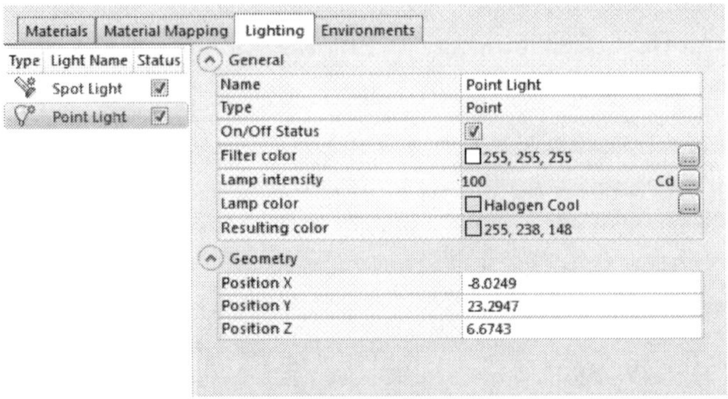

Figure 9-11 *Properties for the Point light in the* ***Properties*** *view area*

11. Click on the browse button displayed in the **Lamp color** edit box; the **Lamp Color** dialog box is displayed.

12. Select the **Halogen Cool** lamp color from the **Standard colors** drop-down list and choose the **OK** button; the selected color is applied to the lamp.

13. Choose the **Light Glyphs** button in the **Autodesk Rendering** window to hide the glyphs from the Scene View .

Specifying Location, Date, and Time of the Scene

In this section, you will set the time, date, and geographic location of the scene.

1. Choose the **Location** button from the Rendering toolbar; the **Geographic Location** dialog box is displayed.

2. In this dialog box, select the **Indiana (East)** option from the **Time Zone** drop-down list and then choose the **OK** button; the dialog box is closed.

3. Now, choose the **Environments** tab and specify date and time in the **Sun Angle Calculator** area of the **Sun** area, as shown in Figure 9-12.

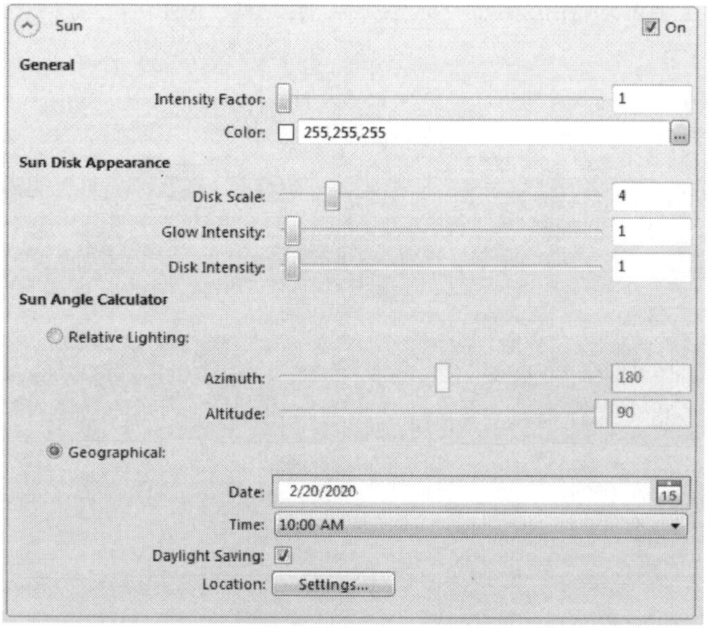

Figure 9-12 *Partial view of the* ***Environments*** *tab*

Figure 9-13 shows the view after changing the location and time.

Figure 9-13 *Model view after changing the location and time*

Rendering the Scene

In this section, you will render the scene with medium quality effects.

1. Close the **Autodesk Rendering** window and select the **Medium Quality** option from **Render > Interactive Ray Trace > Ray Trace** drop-down list.

2. Choose the **Ray Trace** button from the **Interactive Ray Trace** panel of the **Render** tab; the rendering process starts. Figure 9-14 shows the rendered model.

Figure 9-14 *View after rendering*

Similarly, select other rendering options to render the scene and compare the results. Figures 9-15 and 9-16 show a scene rendered using the **Coffee Break Rendering** and **Lunch Break Rendering** options respectively.

Saving the Rendered Scene as an Image

In this section, you will save the rendered scene as a JPEG image.

1. Choose the **Image** button from the **Export** panel of the **Render** tab; the **Save As** dialog box is displayed.

2. Save the image as a JPEG file with the name **c09_navisworks_2024_tut01**.

Figure 9-15 *Scene rendered by using the **Coffee Break Rendering** option*

*Figure 9-16 Scene rendered by using the **Lunch Break Rendering** option*

Saving the Project

In this section, you will save the project.

1. Choose **Save As** from the Application Menu; the **Save As** dialog box is displayed.

2. Browse to the *nws_2024/ c09_nws_2024_tut* folder and enter **c09_navisworks_2024_tut01** in the **File name** edit box. Now, select the **Navisworks File Set (*.nwf)** file format from the **Save as type** drop-down list, and choose the **Save** button.

SELF-EVALUATION TEST

Answer the following questions and then compare them to those given at the end of this chapter:

1. You can invoke the **Autodesk Rendering** window using the __________ drop-down in the __________ panel of the __________ tab.

2. You can produce directional lights by using the __________ option.

3. The __________ button is used to invoke the **Geographic Location** dialog box.

4. The __________ panel in the Autodesk Rendering window displays a list of Autodesk materials library.

5. The __________ option is used to assign material to the selected object.

6. In the **Autodesk Rendering** window, you can modify the materials in the Library panel. (T/F).
7. You can invoke the **Autodesk Rendering** window from the **Review** tab. (T/F)
8. The Planar mapping type is the best mapping type for circular objects. (T/F)
9. In the Box mapping type, material is applied on all the sides of an object. (T/F)
10. There are five tabs in the **Autodesk Rendering** window. (T/F)

REVIEW QUESTIONS

Answer the following questions:

1. Which of the following lights emits lights in all directions from a fixed location?

   a) **Point** b) **Spot**
   c) **Distant** d) **Web**

2. Which of the following mapping types is used for spherical objects?

   a) **Box** b) **Cylinder**
   c) **Spherical** d) **Planar**

3. Which of the following buttons is used to invoke the **Material Editor** dialog box?

   a) **Display** b) **Material Editor**
   c) **Point** d) **Material**

4. Which of the following options is used to produce a quick rendering output?

   a) **Medium Quality** b) **High Quality**
   c) **Low Quality** d) **Ray Trace**

5. Which of the following tabs contains the options for the sun, sky, and exposure effects?

   a) **Lighting** b) **Environments**
   c) **Materials** d) **Render**

6. In Autodesk Rendering, you cannot adjust the appearance of the sun. (T/F)
7. The **High Quality** option takes shortest time for rendering. (T/F)
8. You can specify the geographical location by using the **Geographic Location** dialog box. (T/F)
9. You can save rendered view as an image. (T/F)

10. You can change light intensity and lamp color in the Properties view area of the **Lighting** tab. (T/F)

EXERCISE

Exercise 1 Rendering the House

Download and open the *c09_navisworks_2024_ex1* file from *www.cadsofttech.com*. Render the model using the Autodesk materials and effects. The model elements are saved as selection sets in the **Sets** window. Figure 9-17 shows the model to be used for this exercise. Use the parameters given below to complete this exercise. Figure 9-18 shows the final view of the model after applying materials and effects. **(Expected time: 45 min)**

The following steps are required to complete this exercise:

a. Use Table 9-3 for adding materials.
b. Use Table 9-4 for applying materials.
c. Specify the Time Zone as **(UTC: 06:00) Central America** and time 09:00 AM.
d. Render the scene and export the rendered model as a JPEG image.
e. Save the file as *c09_navisworks_2024_ex01*.

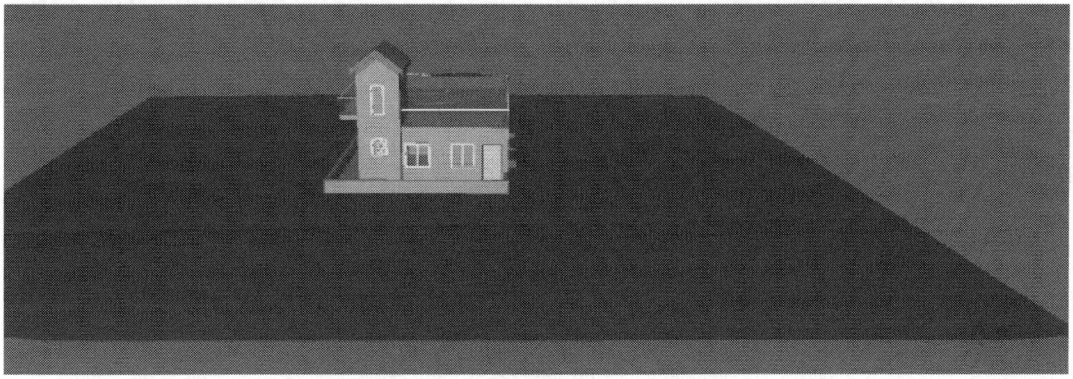

Figure 9-17 *Model displayed in the scene*

Table 9-3 *List of materials to be added to the* ***Document Materials*** *panel*

Materials to be added	Path of the materials in the Library panel
Adobe - Beige	Autodesk Library > Masonry > Brick
Weave - Brown	Autodesk Library > Masonry > Brick
Shake - Handsplit	Autodesk Library > Roofing
Mosaic	Autodesk Library > Flooring > Vinyl
Marble - White	Autodesk Library > Flooring > Stone
Paneling - Yellow Brown	Autodesk Library > Wood > Panels
Yellow Pine - Solid Natural Polished	Autodesk Library > Wood
Clear	Autodesk Library > Glass > Glazing

Table 9-4 *List of objects and the materials assigned to the objects*

Objects in the Sets window	Materials to be Assigned
Outer Walls	Adobe - Beige
Basic Roof 1 & Basic Roof 2	Shake - Handsplit
Roof Floor	Mosaic
Floor	Marble - White
Boundary Walls	Weave - Brown
Window & Door Frames	Yellow Pine - Solid Natural Polished
Door Panel	Panelling - Yellow Brown
Window Glass	Clear

Figure 9-18 *The final view of the model*

Answers to Self-Evaluation Test

1. F, **2.** F, **3.** F, **4.** T, **5.** F, **6. Windows**, **Workspace**, **View**, **7. Spot**, **8. Location**, **9.** Library, **10. Assign to Selection**

Case Studies

Case Study 1

This case study covers the different stages carried to develop the digital model of a residential building in London. In this case study, you will learn about the workflow of Building Information Modelling of a real world project - the Luxury Residential Apartment project. The digital model of this project was developed by The BIM Engineers. These luxury apartments, shown in Figure CS-1, are spread over a span of 9000 square meter and are located in London.

Figure CS-1 *The overall view of the luxury residential apartments*

The building information model was developed by the The BIM Engineers. The scope of this project includes development of Revit Model (Architecture, Structure, and MEP), detecting the clashes within the components, coordination of different disciplines, and generating the construction documentation of the project.

First of all, the structure of the basement level was modeled in Revit. The structure consist of various foundation structures and elements to support the building. Figure CS-2 shows the generated structural model of the building.

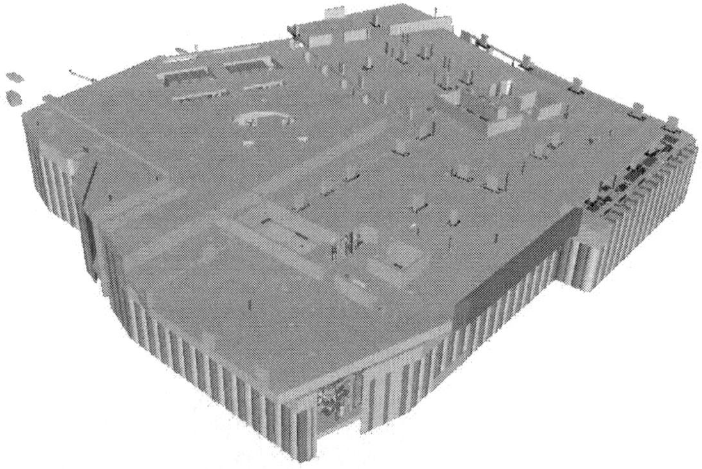

Figure CS-2 *The structural model of the Luxury Resident Apartments at basement level*

After the creation of architectural and structural models, the MEP model of the project was created. The MEP model consists of various elements such as pipes, ducts, fixtures, equipments, and more. Figure CS-3 shows the complete MEP model of the project created in Revit.

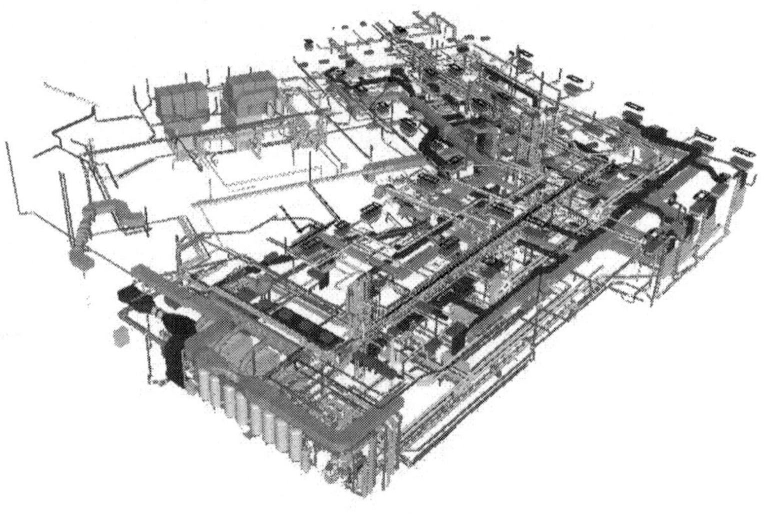

Figure CS-3 *The MEP model of the luxury resident apartment*

During the modeling process various models were generated. Figure CS-4 to Figure CS-10 display various models that were generated for mechanical system.

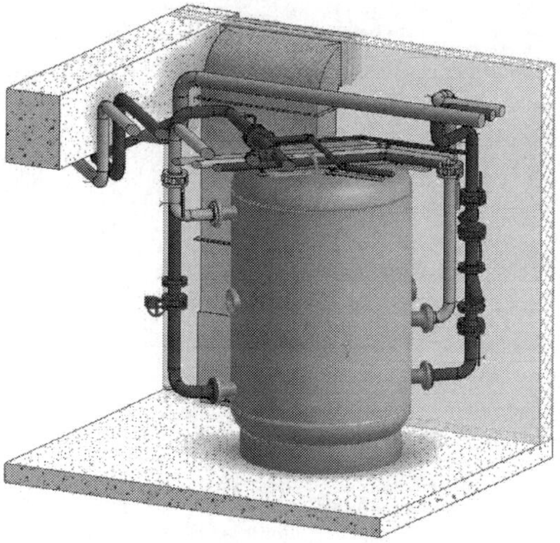

Figure CS-4 *The buffer vessel provision*

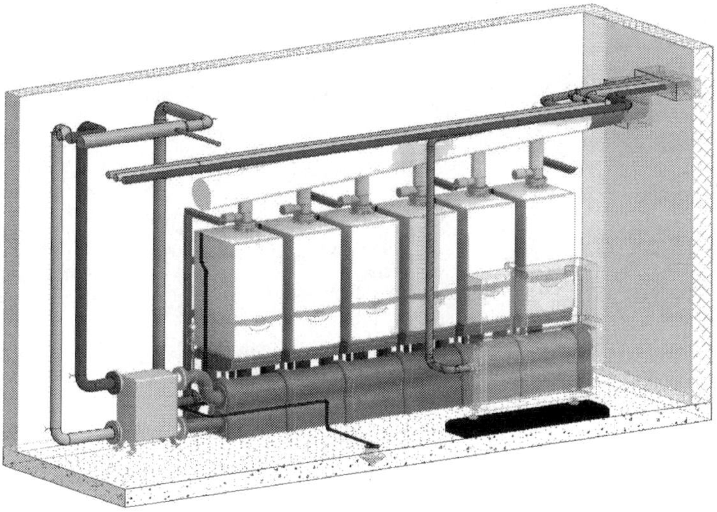

Figure CS-5 *The boiler vessel provisions*

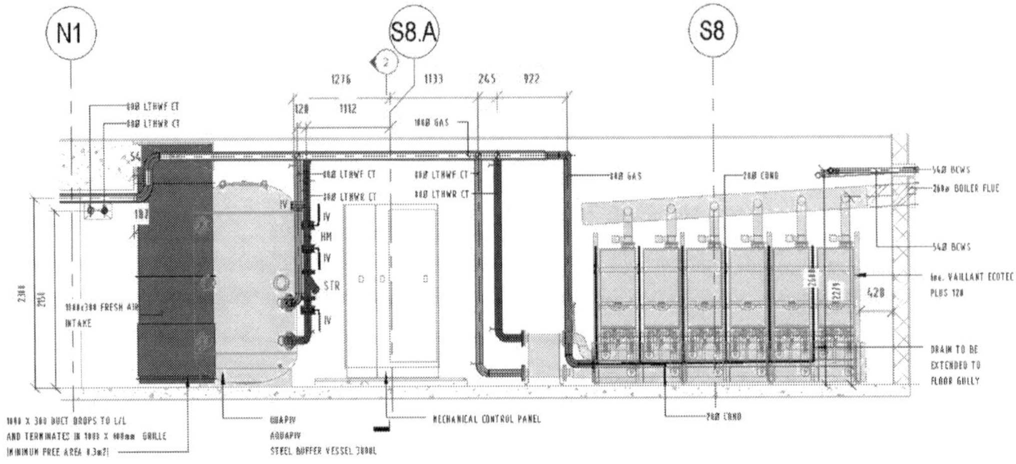

Figure CS-6 *The section details for the boiler and buffer vessel provisions*

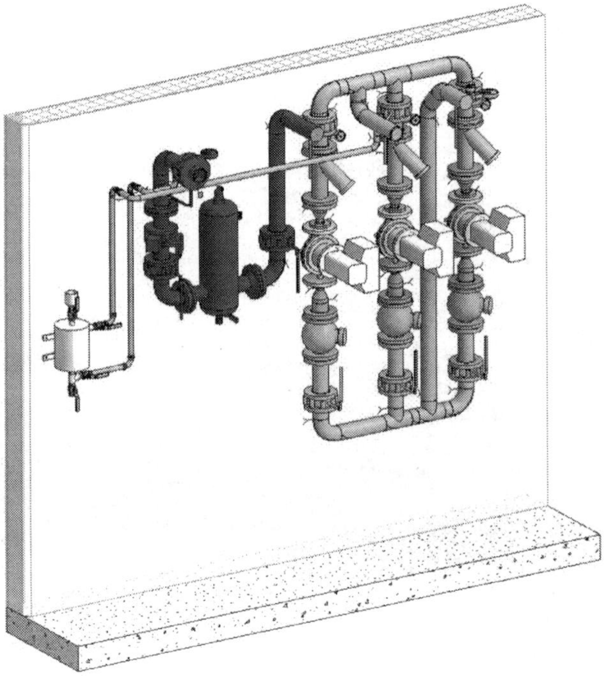

Figure CS-7 *The LTHW pump provisions*

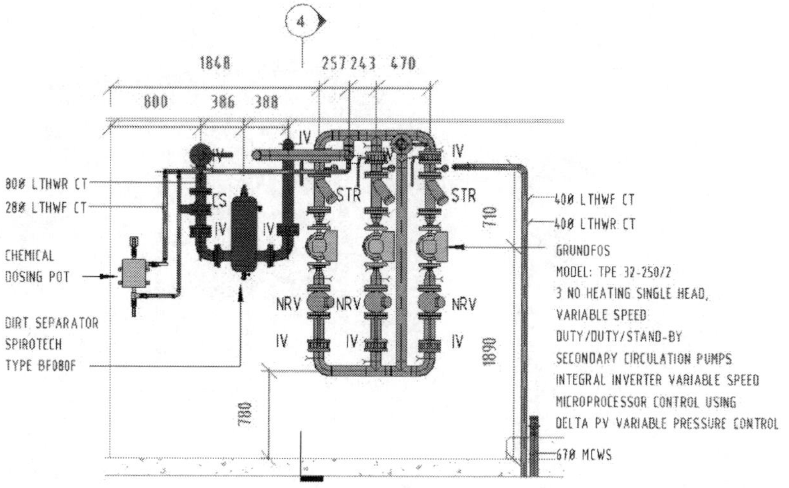

Figure CS-8 *The section details of the LTHW pump*

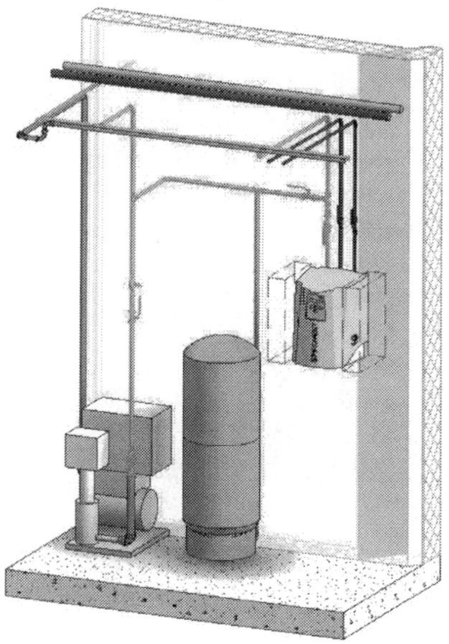

Figure CS-9 *The LTHW pump provisions*

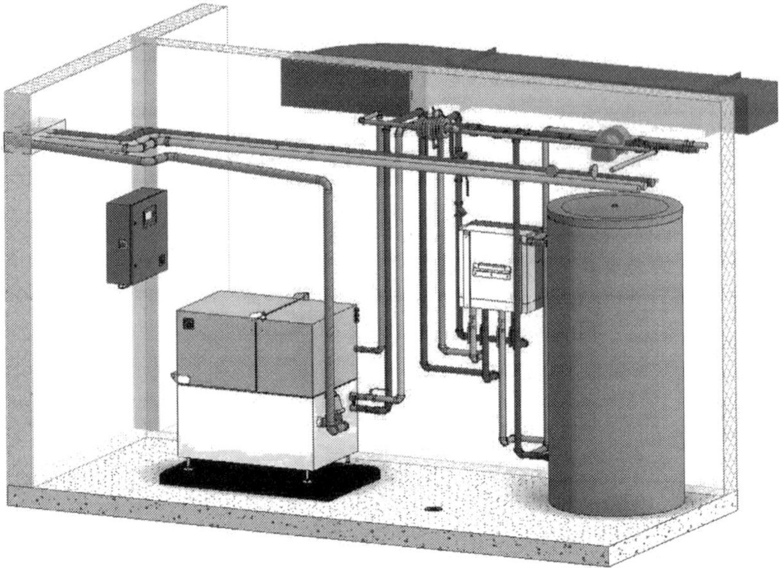

Figure CS-10 *The CHP provisions*

After the MEP model was created, clash detection was performed using Autodesk Navisworks. To perform the clash detection in Navisworks, the MEP model was appended to the Navisworks interface. The clash detection was performed to check the interference betweeen various model elements in the project that were moved during the review process. Figures CS-11 and CS-12 show the clashes between different model elements.

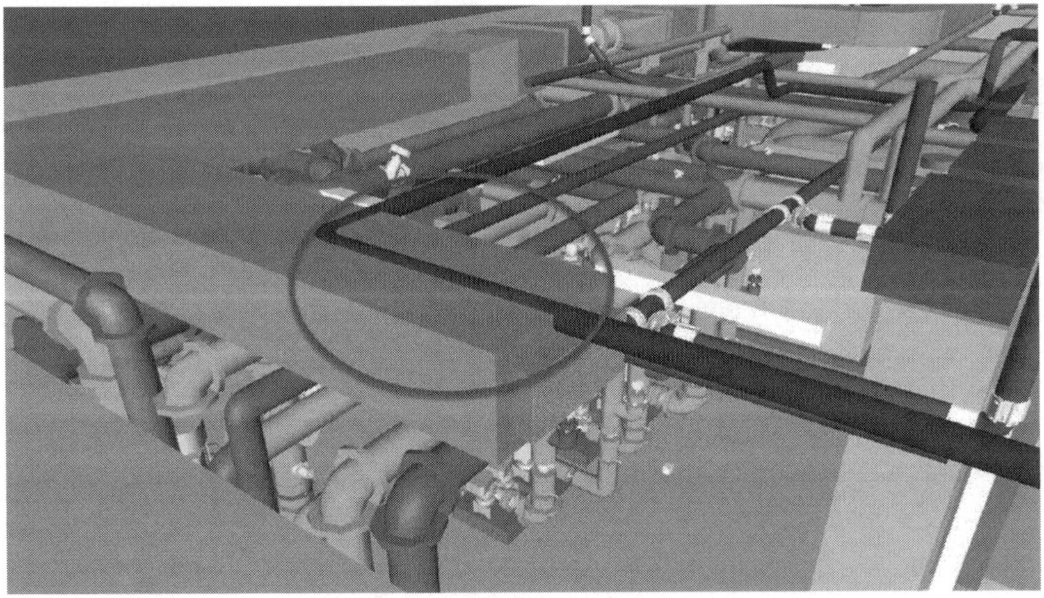

Figure CS-11 *Clash between fire pipes and ducts*

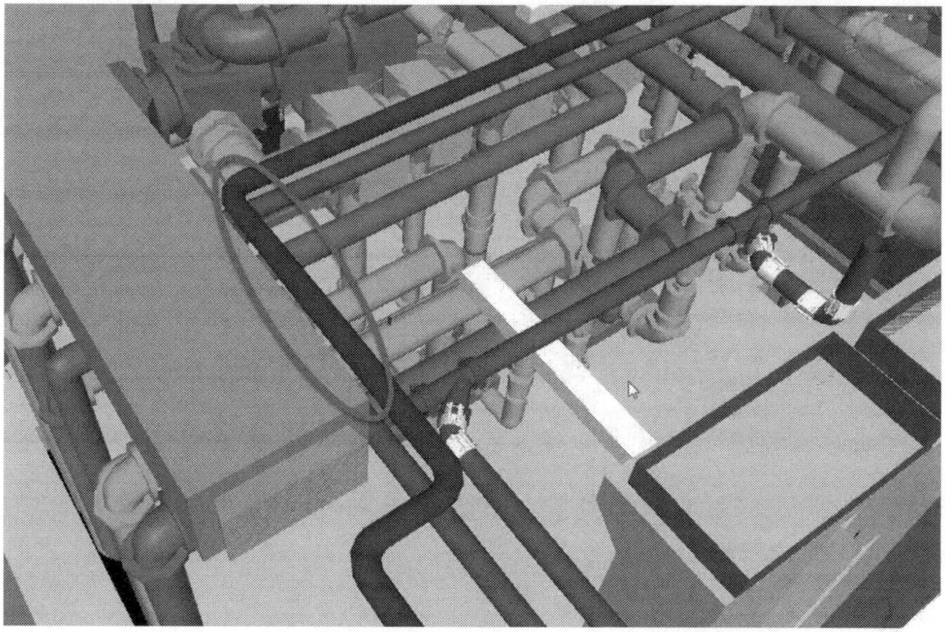

Figure CS-12 *The fire pipe rerouted to resolve the clash*

After the clashes were identified in the model, all interference leading to clashes were removed to make it free of any overlap or error. Further the model was reviewed and navigated for the coordination and presentation purpose. Figure CS-13 to Figure CS-17 display various views of the project for model display.

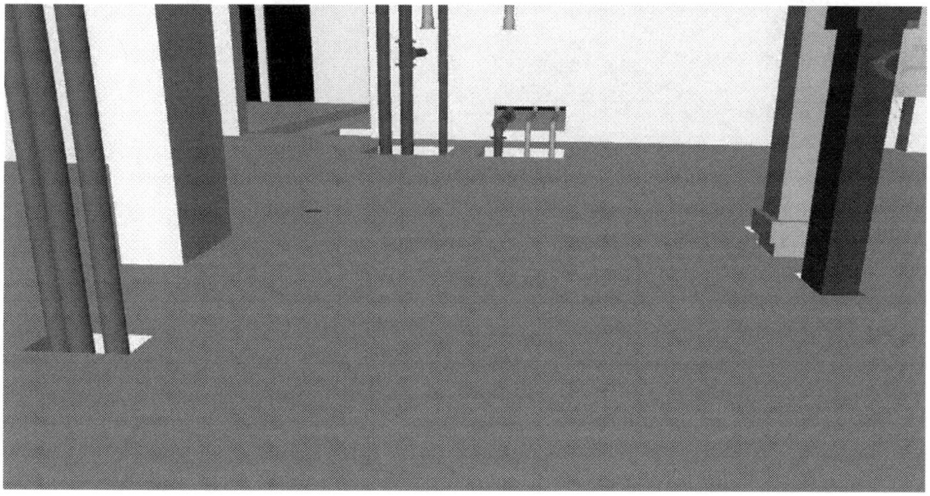

Figure CS-13 *Structural floor and wall openings for MEP services*

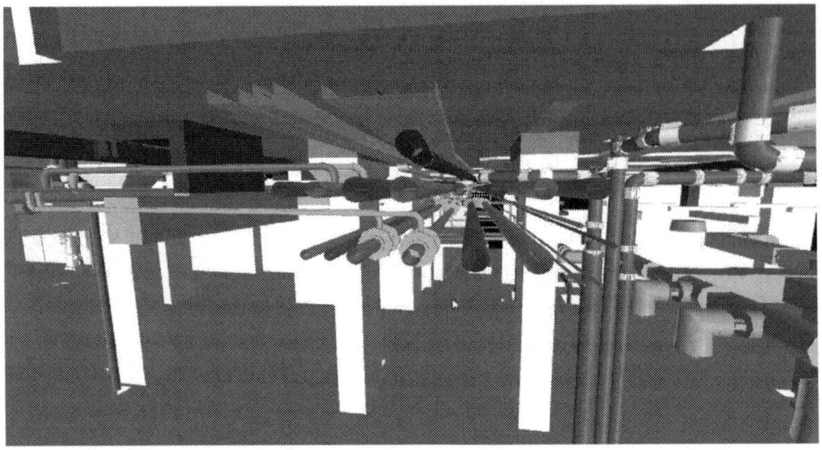

Figure CS-14 *MEP coordination - pipes and tray arrangement*

Figure CS-15 *The boiler plant room*

Figure CS-16 *The chiller plant room*

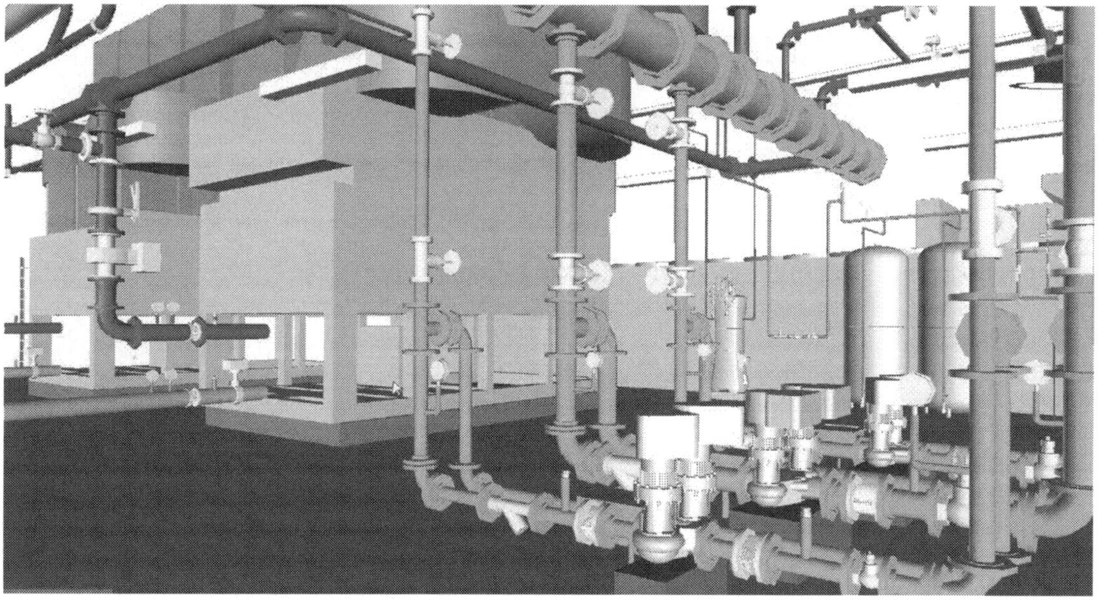

Figure CS-17 *Chiller plant room – equipment, accessories, and pipe connections*

After the coordination process the construction documentation was created. During this, the sheets were created for the final submission to the site. This was done using the Revit software. Figure CS-18 and Figure CS-19 show the sheets published for the project.

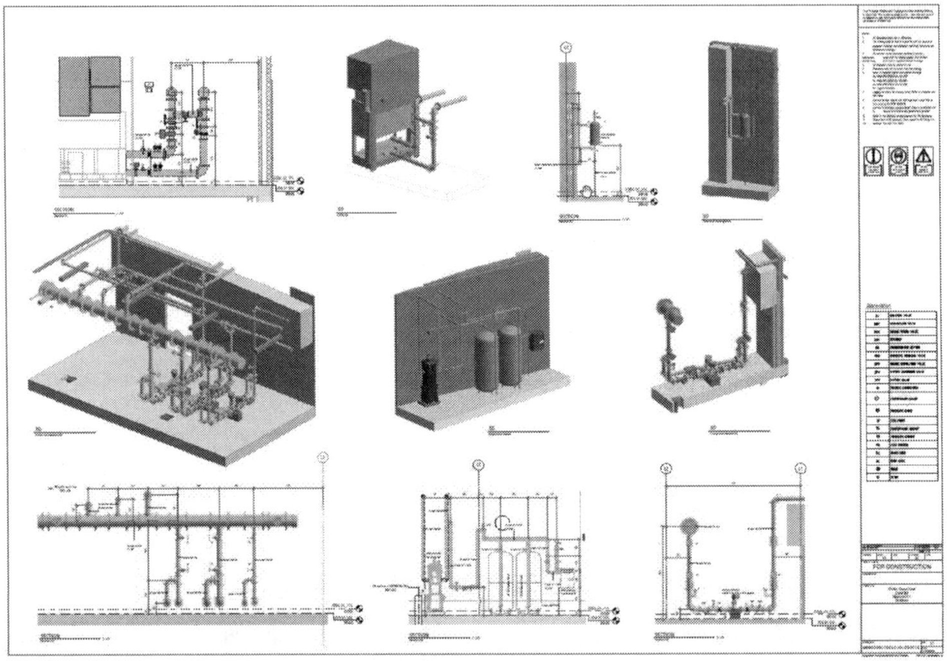

Figure CS-18 *Chiller plant room - Sheet drawing presentation of detailed sections and 3D views*

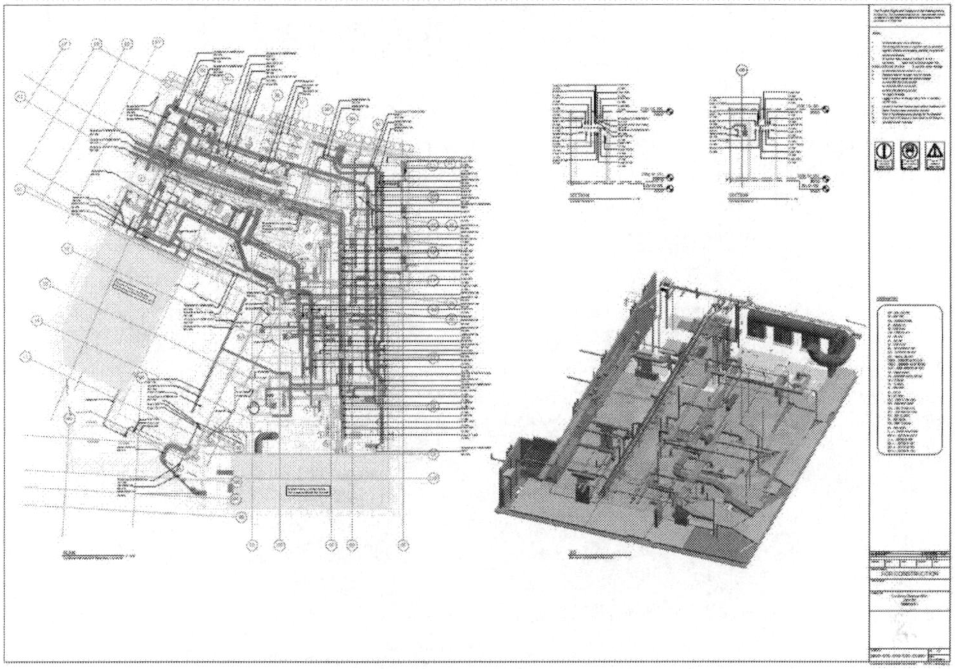

Figure CS-19 *Combined services drawing presentation*

Case Study 2

This case study covers the different stages carried to develop the digital model of a mass rapid station in Kuala Lumpur. The digital model of the Mass Rapid Transit project is developed in Revit and Navisworks. The project involved modeling and documentation in Revit, Clash detection and Coordination in Navisworks. The digital model was completed in 6 months. Figure CS-20 to Figure CS-29 show the progress of the project.

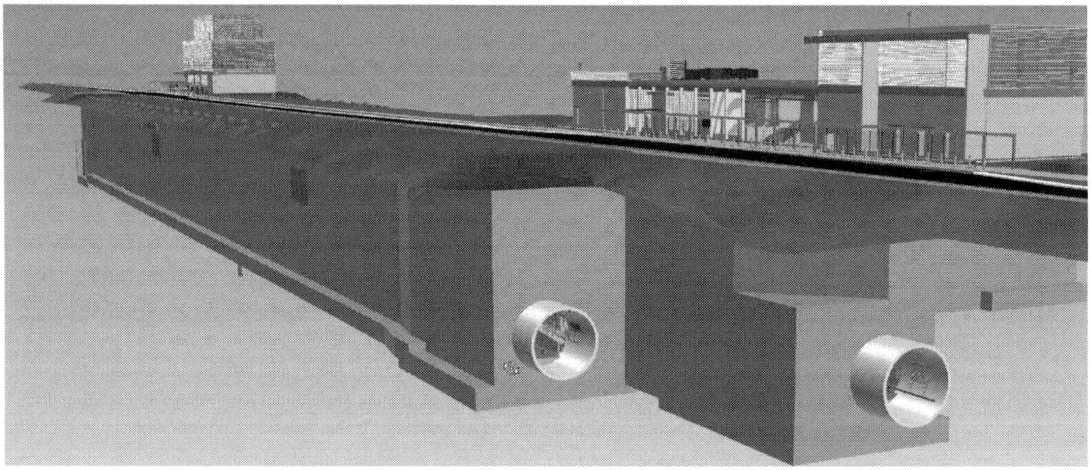

Figure CS-20 *Overall Station view– Architecture, Structure, and MEP*

Figure CS-21 *Overall Station - MEP view*

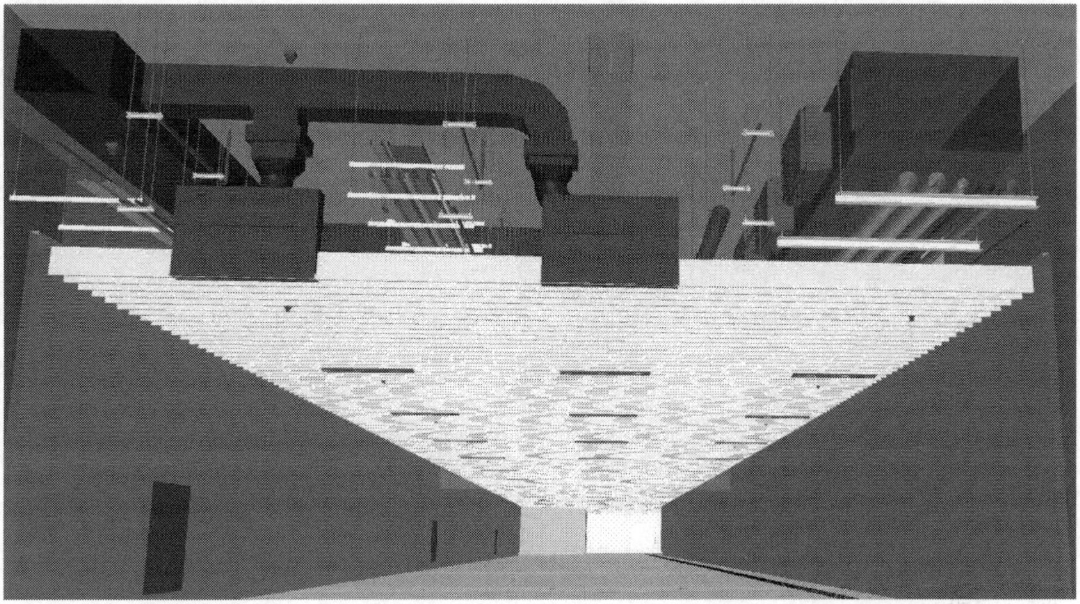

Figure CS-22 *The supply and return air ducts with dampers and diffusers*

Figure CS-23 *Electrical services – cable trays, lighting fixtures, and panels*

Figure CS-24 *Chiller plant room*

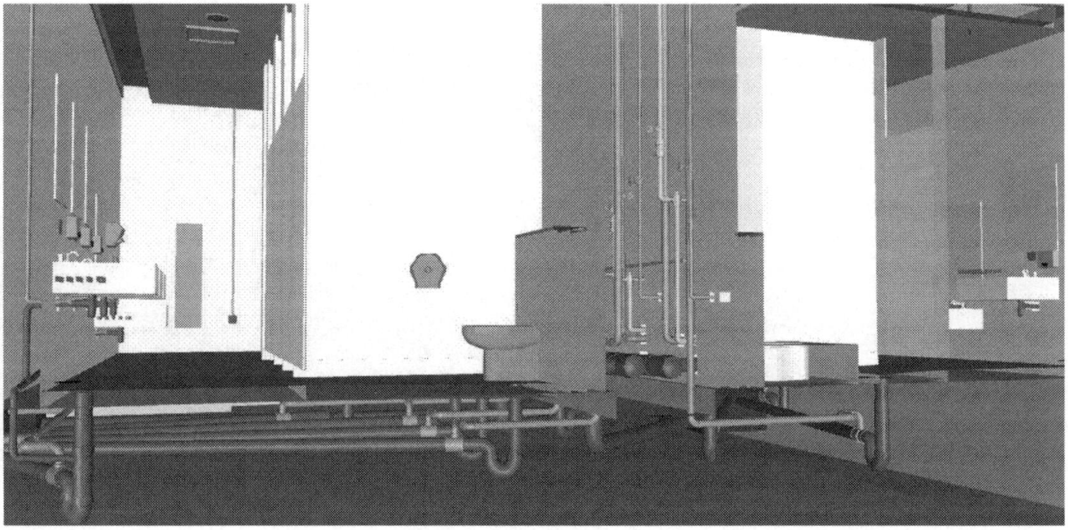

Figure CS-25 *Sanitary system*

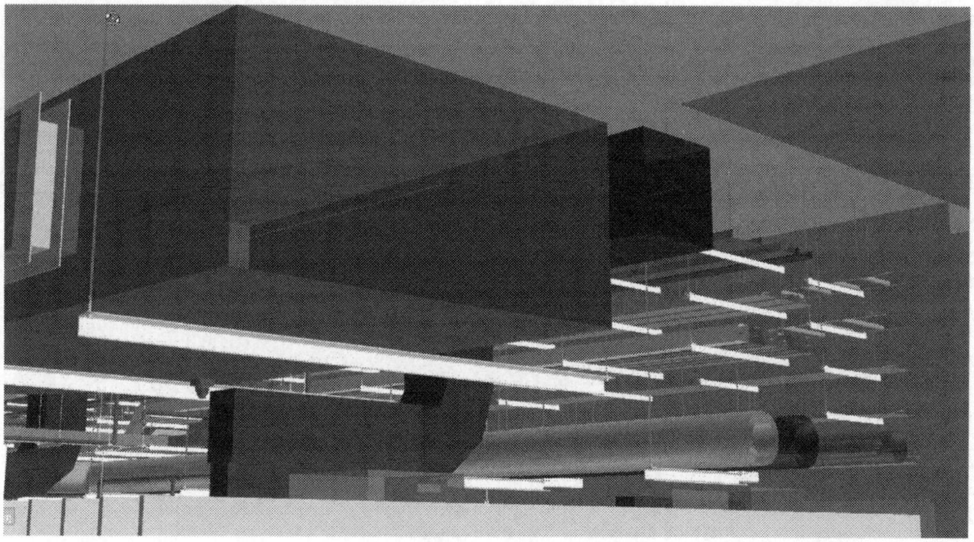

Figure CS-26 *Hanger support*

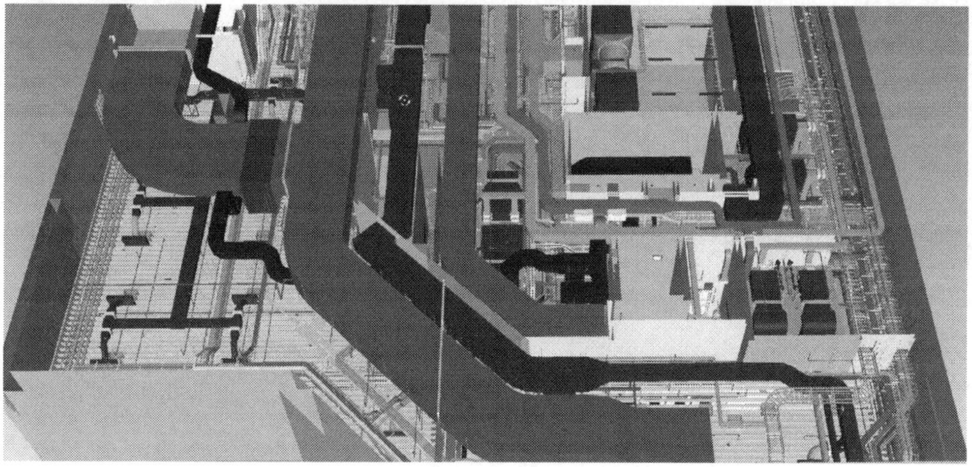

Figure CS-27 *Coordinated MEP services with hanger support*

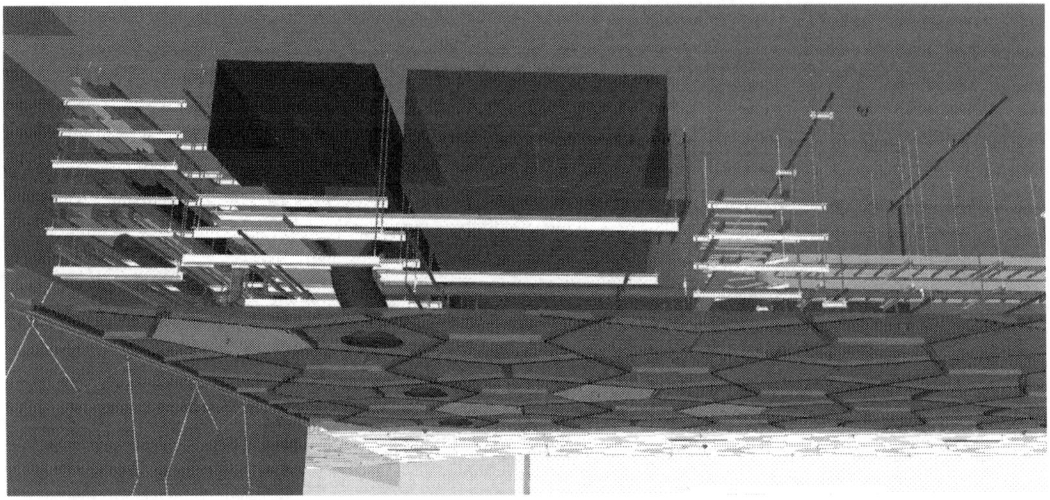

Figure CS-28 *Coordinated MEP services*

Figure CS-29 *Coordinated MEP Services with hanger support*

Index

A

Accumulate tool 3-3
Angle tool 3-3
Animator window 6-2
Area tool 3-3

B

Batch utility 1-10
Box mode 4-3

C

Camera 2-4
Clear tool 3-3
Comments 3-4

D

Document Materials panel 9-7

E

Enabling Scripts 6-5
External Database 3-3
Exposure 9-3

F

Find Items 3-2
Fly tool 2-4

G

Gizmo 3-5

H

Head-Up Display 2-2

I

Item Catalog 7-2

L

Library panel 9-3

M

Material Mapping 9-3
Measure tools 3-3
Model Takeoff 7-4
Markup tools 3-4

N

Navisworks Workspaces 1-11
New Master Resource dialog box 7-14

O

Object Properties 3-3
Orientations 2-2

P

Parallel tool 3-4
Perpendicular tool 3-4
Planes Mode 4-2
Point Line tool 3-3
Point to Multiple Points tool 3-3

Q

Quick Line 7-6

R

Run Test 8-4

S

Scripter 6-4, 6-5
Select All 3-3
Selection Inspector 3-3
Selection Resolution 3-2
Selection Tree 3-2
Settings tab 9-3
Soft Clashing 8-4
Sun 9-3

T

Time-Based Clash Test 8-4
TimeLiner Costs 5-4

U

V

ViewCube 1-6, 2-3
Virtual Takeoff 7-3, 7-4

W

Walk tool 2-4

X

X Axis tool 3-4

Y

Y Axis tool 3-4

Z

Z Axis tool 3-4

Other Publications by CADSoft Technologies

The following is the list of some of the publications by CADSoft Technologies. Please visit *www.cadsofttech.com* for the complete listing.

Revit Architecture for Novices Textbooks
- Revit Architecture 2020 for Novices: Learn By Doing
- Revit Architecture 2019 for Novices: Learn By Doing
- Revit Architecture 2018 for Novices: Learn By Doing

Revit Structure for Novices Textbooks
- Revit Structure 2024 for Novices: Learn By Doing
- Revit Structure 2023 for Novices: Learn By Doing
- Revit Structure 2020 for Novices: Learn By Doing

Revit MEP for Novices Textbook
- Revit MEP 2019 for Novices: Learn By Doing

AutoCAD Civil 3D for Novices Textbook
- AutoCAD Civil 3D 2023 for Novices: Learn By Doing

Autodesk 3dx Max for Novices Textbooks
- Autodesk 3ds Max 2020 for Novices: Learn By Doing
- Autodesk 3ds Max 2019 for Novices: Learn By Doing

Autodesk Maya for Novices Textbook
- Autodesk Maya 2019 for Novices: Learn By Doing

Pixologic ZBrush for Novices Textbook
- Pixologic ZBrush 2018 for Novices: Learn By Doing

MAXON CINEMA 4D for Novices Textbook
- MAXON CINEMA 4D R20 Studio for Novices: Learn By Doing

AutoCAD for Novices Textbooks
- AutoCAD 2020 for Novices: Learn By Doing
- AutoCAD 2019 for Novices: Learn By Doing
- AutoCAD 2018 for Novices: Learn By Doing

Inventor Professional for Novices Textbook
- Inventor 2020 for Novices: Learn By Doing

Siemens NX for Novices Textbook
- Siemens NX 2019 for Novices: (Continous Release) Learn By Doing

Creo Parametric for Novices Textbooks

- Creo Parametric 6.0 for Novices: Learn By Doing
- Creo Parametric 5.0 for Novices: Learn By Doing

SOLIDWORKS for Novices Textbook

- SOLIDWORKS 2018 for Novices: Learn By Doing

Made in United States
Orlando, FL
04 November 2026

90200106R00133